The ROLE of the NUCLEUS of the SOLITARY TRACT in GUSTATORY PROCESSING

, Ph.D., Professor of Anatomy and Neurobiology,
University of Tennessee Health Sciences Center

The Superior Colliculus: New Approaches for Studying Sensorimotor Integration
William C. Hall, Ph.D., Department of Neuroscience, Duke University
Adonis Moschovakis, Ph.D., Institute of Applied and Computational Mathematics, Crete

New Concepts in Cerebral Ischemia
Rick C. S. Lin, Ph.D., Professor of Anatomy, University of Mississippi Medical Center

DNA Arrays: Technologies and Experimental Strategies
Elena Grigorenko, Ph.D., Technology Development Group, Millennium Pharmaceuticals

Methods for Alcohol-Related Neuroscience Research
Yuan Liu, Ph.D., National Institute of Neurological Disorders and Stroke, National Institutes
of Health
David M. Lovinger, Ph.D., Laboratory of Integrative Neuroscience, NIAAA

Optical Imaging of Brain Function
Ron Frostig, Ph.D., Associate Professor/Department of Psychobiology,
University of California, Irvine

Primate Audition: Behavior and Neurobiology
Asif A. Ghazanfar, Ph.D., Primate Cognitive Neuroscience Lab, Harvard University

Methods in Drug Abuse Research: Cellular and Circuit Level Analyses
Dr. Barry D. Waterhouse, Ph.D., MCP-Hahnemann University

Functional and Neural Mechanisms of Interval Timing
Warren H. Meck, Ph.D., Professor of Psychology, Duke University

Biomedical Imaging in Experimental Neuroscience
Nick Van Bruggen, Ph.D., Department of Neuroscience Genentech, Inc., South San Francisco
Timothy P.L. Roberts, Ph.D., Associate Professor, University of Toronto

The Primate Visual System
John H. Kaas, Department of Psychology, Vanderbilt University
Christine Collins, Department of Psychology, Vanderbilt University

Neurosteroid Effects in the Central Nervous System
Sheryl S. Smith, Ph.D., Department of Physiology, SUNY Health Science Center

Modern Neurosurgery: Clinical Translation of Neuroscience Advances
Dennis A. Turner, Department of Surgery, Division of Neurosurgery, Duke University
Medical Center

Sleep: Circuits and Functions
Pierre-Hervé Luoou, Université Claude Bernard Lyon I, Lyon, France

Methods in Insect Sensory Neuroscience
Thomas A. Christensen, Arizona Research Laboratories, Division of Neurobiology, University
of Arizona, Tucson, AZ

The ROLE of the NUCLEUS of the SOLITARY TRACT in GUSTATORY PROCESSING

Edited by

Robert M. Bradley

University of Michigan
Ann Arbor, Michigan

CRC Press
Taylor & Francis Group
Boca Raton London New York

CRC Press is an imprint of the
Taylor & Francis Group, an **informa** business
A TAYLOR & FRANCIS BOOK

CRC Press
Taylor & Francis Group
6000 Broken Sound Parkway NW, Suite 300
Boca Raton, FL 33487-2742

First issued in paperback 2019

ISBN-13: 978-0-8493-4200-4 (hbk)
ISBN-13: 978-0-367-39019-8 (pbk)

Library of Congress Cataloging-in-Publication Data

The role of the nucleus of the solitary tract in gustatory processing / Robert M. Bradley, editor.
    p. cm. -- (Frontiers in neuroscience ; 31)
    Includes bibliographical references (p.   ).
  ISBN 0-8493-4200-7 (alk. paper)
  1. Solitary nucleus. 2. Taste. I. Bradley, Robert M. (Robert Martin), 1939- II. Series: Frontiers in neuroscience (Boca Raton, Fla.)

QP377.R65 2006
612.8'7--dc22   2006047539

Visit the Taylor & Francis Web site at
http://www.taylorandfrancis.com

and the CRC Press Web site at
http://www.crcpress.com

Table of Contents

Series Preface

The Frontiers in Neuroscience series presents the insights of experts on emerging experimental techniques and theoretical concepts that are or will be at the vanguard of neuroscience. Books in the series cover topics ranging from methods to investigate apoptosis to modern techniques for neural ensemble recordings in behaving animals. The series also covers new and exciting multidisciplinary areas of brain research, such as computational neuroscience and neuroengineering, and describes breakthroughs in fields like insect sensory neuroscience, primate audition, and biomedical engineering. The goal is for this series to be the reference that every neuroscientist uses in order to get acquainted with new methodologies in brain research. These books can be given to graduate students and postdoctoral fellows when they are looking for guidance to start a new line of research.

Each book is edited by an expert and consists of chapters written by the leaders in a particular field. Books are richly illustrated and contain comprehensive bibliographies. Chapters provide substantial background material relevant to the particular subject. Hence, they are not the usual type of method books. They contain detailed "tricks of the trade" and information as to where these methods can be safely applied. In addition, they include information about where to buy equipment and web sites helpful in solving both practical and theoretical problems. Finally, they present detailed discussions of the present knowledge of the field and where it should go.

We hope that, as the volumes become available, the effort put in by us, the publisher, the book editors, and the individual authors will contribute to the further development of brain research. The extent to which we achieve this goal will be determined by the utility of these books.

Sidney A. Simon, Ph.D.
Miguel A.L. Nicolelis, M.D., Ph.D.
Series Editors

Preface

Recent discoveries of G-protein coupled membrane receptors in taste cells have resulted in a reexamination of the mechanisms of taste transduction and how different taste qualities are encoded. In particular, sweet, bitter, and amino acid (umami) receptors have been identified and apparently localized to separate cells within the taste bud. Investigators now describe taste buds as having sweet cells, amino acid cells, and bitter cells. The existence of salt and sour cells is usually not mentioned. The assumption increasingly made is that quality coding in the taste system is hardwired with direct connections (so-called labeled-line) between taste cells that express a particular receptor type and the central nervous system area that is responsible for perceptual behavior, the concept being that there are separate taste pathways for preference and aversive behaviors. Many of the previous investigations of taste coding argued against this labeled-line explanation based on taste bud cell turnover and peripheral and central nervous system neurophysiological investigations. More to the point, the labeled-line hypothesis assumes that the relays in the central taste pathway have a minimal role in sensory processing, merely serving to connect the different ascending tracts of the gustatory pathway. However, much research has established that although specificity might exist to some extent at the taste periphery, it is gradually degraded at each relay nucleus in the ascending taste pathway.

The first relay in the taste pathway is the nucleus of the solitary tract, a gateway to central taste processing. However, the neurobiology of the solitary tract nucleus and the other central nervous system taste relays has received relatively little attention. Investigators who examined the anatomical and neurophysiological properties of the central taste relay nuclei have demonstrated that afferent taste information changes as it passes through each relay. For example, central neurons that respond to chemicals that taste sweet or bitter also respond to salts and acids, demonstrating that convergence of information occurs. Although early research seemed to indicate that the nucleus of the solitary tract is a relatively simple relay, more recent investigations using a variety of techniques have revealed that this nucleus is quite complex; besides being part of the ascending taste pathway, the nucleus also serves to connect to several brainstem sites responsible for a number of important taste-initiated reflex functions.

A principal motivation for assembling this book was to bring together in one volume the expertise of a number of investigators who have studied the nucleus of the solitary tract and have contributed substantially to the current knowledge of the anatomy, physiology, and developmental biology of the solitary nucleus.

My aim was to provide the reader with information that has never before been gathered in one place to serve as a reference and hopefully motivate others to bring new approaches to advance knowledge about the central processing of gustatory information. As the reader will find, much has been accomplished, but the time is ripe for progress in revealing what the solitary nucleus does with the gustatory information delivered to its afferent portal.

Editor

Robert M. Bradley is a professor in the Department of Biologic and Materials Sciences in the School of Dentistry of the University of Michigan. He earned a B.D.S. from the University of London, an L.D.S. from the Royal College of Surgeons of England, and an M.S.D. from the University of Washington. Dr. Bradley began his investigations of the gustatory system under Lloyd M. Beidler, which resulted in a Ph.D. Subsequently, he investigated the development of the taste system during his postdoctoral studies at the Nuffield Institute of Oxford University in England. In 1972, he joined the faculty of the School of Dentistry at the University of Michigan. Dr. Bradley has continued to investigate the gustatory system using both anatomical and electrophysiological techniques. For the last 20 years, he has concentrated on the brainstem gustatory nucleus.

Dr. Bradley is a member of the Society for Neuroscience and the American Physiological Society. He is a founding member of the Association for Chemoreception Sciences and was awarded the Max Mozell Award for Outstanding Achievement in the Chemical Senses in 2003. He has authored or coauthored over 150 journal articles and book chapters on the chemical senses, and 2 books on oral physiology.

List of Contributors

Dr. Robert M. Bradley
Department of Biologic and Materials
Science
School of Dentistry
University of Michigan
Ann Arbor, Michigan

Dr. David L. Hill
Department of Psychology
University of Virginia
Charlottesville, Virginia

Dr. Miwon Kim
Center for Nursing Research
School of Nursing
Chonnam National University
Gwangju, Republic of Korea

Dr. Michael S. King
Department of Biology
Stetson University
DeLand, Florida

Dr. Christian H. Lemon
Department of Anatomy and
Neurobiology
University of Tennessee Health Science
Center
Memphis, Tennessee

Dr. Olivia L. May
Department of Biologic and Materials
Science
School of Dentistry
University of Michigan
Ann Arbor, Michigan

Dr. David V. Smith
Department of Anatomy and
Neurobiology
University of Tennessee Health Science
Center
Memphis, Tennessee

The ROLE of the NUCLEUS of the SOLITARY TRACT in GUSTATORY PROCESSING

1 Historical Perspectives

Robert M. Bradley

CONTENTS

1.1 BASIC NEUROANATOMY OF THE TERMINATION OF THE AFFERENT INPUT TO THE NUCLEUS OF THE SOLITARY TRACT

Dating from the 1960s, the basic neuroanatomy of the gustatory relay nucleus in the medulla was established, and all further research relied on this foundation. Specifically, afferent fibers innervating taste buds in the oral cavity, pharynx, and larynx synapse with second-order neurons in the nucleus of the solitary tract (NST). These afferent fibers travel in the facial (VII), glossopharyngeal (IX), and vagus (X) nerves. Their termination in the NST is topographical, the anterior oral cavity terminating in the rostral extent of the NST, the posterior oral cavity terminating more caudally, and the pharyngeal and laryngeal receptive field even more caudally (Figure 1.1). All current studies of the NST assume this basic arrangement of the brainstem taste relay. However, much research was required before these pathways and termination patterns were firmly established. As recently as 1950, Patton stated in a review that "the location of the secondary neurons of the taste pathway is undetermined,"[1] yet 10 years later students in Carl Pfaffmann's laboratory were recording from taste responsive neurons in the NST (see Section 1.5).

In this chapter, the long history of the anatomical research that resulted in the current understanding of the neuroanatomy of the NST will be reviewed. Many investigators have contributed to the current knowledge of the rostral NST (rNST), and their successes have often relied not only on their powers of observation but

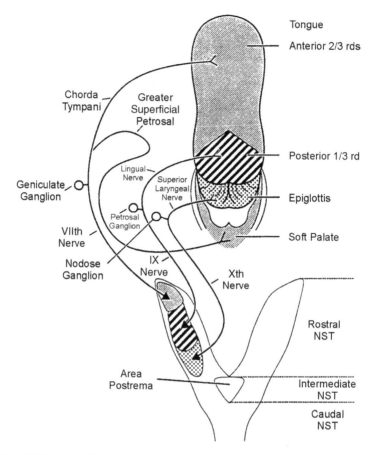

FIGURE 1.1 Diagram of the projection of the afferent sensory innervation of the tongue, soft palate, and larynx onto the nucleus of the solitary tract in the medulla. (Derived from Loewy, A. D., Central autonomic pathways, in *Central Regulation of Autonomic Functions*, Loewy, A.D. and Spyer, K.M. (Eds.), Oxford University Press, Oxford, 1990. With permission.)

also on the techniques available to them at the time. An understanding of the many anatomical and clinical studies that preceded our current understanding of the neurobiology of the NST provides the foundation for the content of all the subsequent chapters in this book.

1.2 DISCOVERY OF THE SOLITARY TRACT

The earliest findings leading to the discovery of the gustatory brainstem relay nucleus was the identification in cross sections of the medulla of the solitary tract (ST). Stilling[2] and Clarke[3,4] provide the earliest descriptions of a longitudinal fiber bundle in the medulla. In describing the glossopharyngeal nerve, Clarke in 1858[3] stated that the "nerve passes inward and backwards, in two or three bundles, through the gelatinous substance and across the arciform fibres. On reaching the

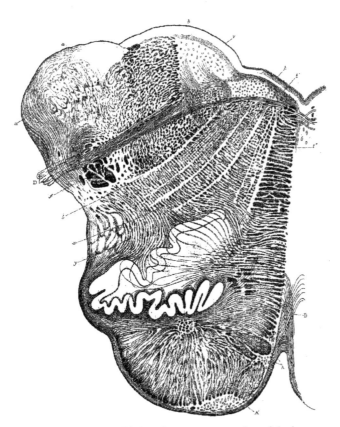

FIGURE 1.2 An 1858 drawing by Clarke of a transverse section of the human medulla at the level of the roots of the vagus nerve (D). Although the vagal fibers are diagrammed to terminate in the caudal nucleus of the solitary tract, the nucleus is not labeled and some of the vagal fibers are diagrammed to cross the midline. Despite these errors, this diagram is remarkable for its time. (From Clarke, J. L., *Phil. Trans. Royal Soc. London*, 148, Plate XVII, 1858.)

group of *longitudinal fascicles*, which lie at the extremity of the vagal nucleus, its outer portion separates in a brush-like manner into many smaller bundles, which subdivide the fasciculi into a corresponding number of parts. Many of its fibers at this point appear to become longitudinal." Figure 1.2 is taken from Clarke's publication, and although the entering glossopharyngeal nerve fibers are labeled (D in Figure 1.2), the NST and ST are clearly drawn, they are not labeled. In a later publication, Clarke[4] referred to the ST as the "slender longitudinal column," which has "some kind of important connection with the vagal and glossopharyngeal nuclei." Clarke cut an oblique longitudinal section of the human medulla in the plane of the glossopharyngeal nerve, passing through the slender column *nn* and the glossopharyngeal nerve *t*. Some of the roots of the nerve (*g*) are distinctly seen to enter the slender column (*n*) and run *down* the medulla (Figure 1.3).

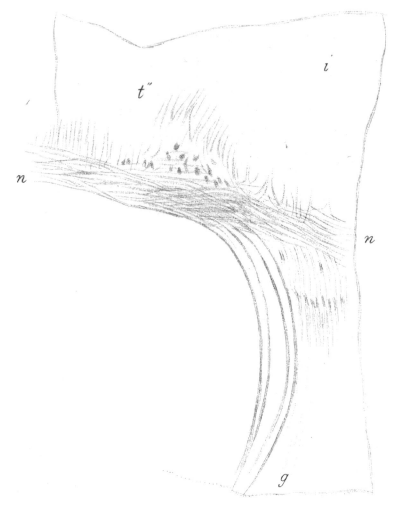

FIGURE 1.3 An 1868 drawing by Clarke of a longitudinal section of part of the human medulla. *t"* is the glossopharyngeal nucleus, *g* the roots of the glossopharyngeal nerve, and *nn*, the slender longitudinal column, i.e., the solitary tract. In the text of the publication, the "slender longitudinal column" is stated to have an important connection with both the vagal and glossopharyngeal nuclei. Clarke also states that some of the roots of the glossopharyngeal nerve are "distinctly seen to enter the slender column and run with it *down* the medulla." (From Clarke, J. L., *Phil. Trans. Royal Soc. London*, 158, Plate X, 1868.)

1.3   DISCOVERY OF THE AFFERENT SENSORY CONTRIBUTION TO THE SOLITARY TRACT

Clarke provided the first description of the contribution of the glossopharyngeal nerve to the ST. Meynert[5] is the earliest author to name the tract the solitäres Bündel (solitary tract), a term that Duval[6] preferred to use (bandelette solitaire) because it is a purely anatomical name that does not predict its connections or functional role.

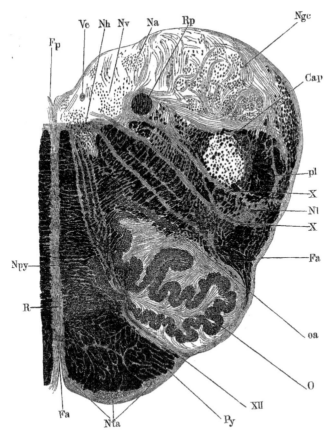

FIGURE 1.4 Transverse section of the medulla at the inferior end of the 4th ventricle. The ST is clearly stained and labeled Rp, the Respirationsbündel. The respiratory role of the ST suggested by this nomenclature derives from the early discovery of its association with the vagus and glossopharyngeal nerves. Because afferent input conveyed to the caudal ST is important in respiratory reflexes,[38] the early nomenclature is partially correct. However, using a name that has no functional connotation to a tract that carries afferent information associated with multiple functions avoids potential confusion. (From Krause, C. F. T., *Handbuch der Menschlichen Anat.*, 1876, p. 410.)

Krause,[7] on the other hand, referred to the ST as the Respirationsbündel (Figure 1.4), presumably based on speculation of its possible function. However, by the late 1800s, the name of the solitary tract was established and consistently used by all authors in publications, although in various translations (solitäres Bündel, bandelette solitaire, fasiculus solitarius, and solitary tract).

The earlier authors seemed to agree that the afferent fibers contributing to the ST derived solely from the IXth nerve.[8,9] However, Dees[10] stated that the Xth nerve also contributed. According to Koelliker,[11] Cramer,[12] and Hiss,[13] the ST derives its fibers from both the IXth and Xth nerves. Koelliker also states that surrounding this bundle is a considerable amount of gelatinous substance, into which axis cylinders and their collaterals terminate. Within the gelatinous substance are cells that send their processes into the reticular formation, and thus the ST is associated with other parts of the brain. This is the first description of the NST, at this time called the dorsal sensory nucleus of the glossopharyngeal nerve.[14] Some confusion existed as to whether the dorsal sensory nucleus of the glossopharyngeal nerve was sensory or motor. Koelliker,[11] Ramón Y Cajal,[15] and Van Gehuchten[16] all agreed that the sensory nucleus of the glossopharyngeal nerve was sensory, whereas other investigators[10] maintained that the nucleus was motor. Based on the results of nerve sectioning and by studying a case in which a tumor had destroyed the trigeminal and glossopharyngeal roots, Bruce[17] reported that the "fasiculus solitarius is an afferent nerve belonging in its upper part to the glossopharyngeal nerve, and in its middle and lower part mainly to the vagus nerve."

By 1900, the contributions of the IXth and Xth nerves to the ST were firmly established. The termination of the sensory root of the VIIth nerve, however, was less well understood and according to some authors did not contribute to the ST (Figure 1.5). Part of the problem centered on the controversy regarding the nature of the facial nerve. Despite the fact that Lussana in 1869 (reference cited in Schwartz and Weddell[18]) had, on the basis of clinical and experimental observations, described the pathway for transmission of gustatory sensation from the anterior two thirds of the tongue via the chorda tympani, other investigators reached different conclusions. As detailed by Dixon,[19] many investigators considered the facial nerve to be purely motor. Moreover, although the chorda tympani innervation of the tongue was well established, it was frequently assumed to reach the brainstem via the trigeminal nerve.

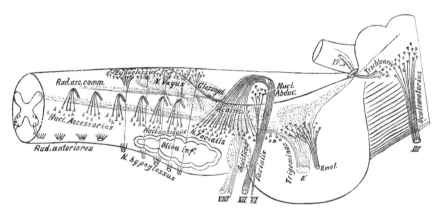

FIGURE 1.5 Early diagram summarizing the brainstem connections of the cranial nerves IV–XII. The motor nerves are black, and the sensory nerves are stippled. The vagus and glossopharyngeal nerves are diagrammed to terminate in a common nucleus. The facial nerve (VII) is purely motor. (From Bekhterev, V. M., *Die Leitungsbahnen im Gehirn und Rückenmark,* Georgi, Leipzig 1899, p.144.)

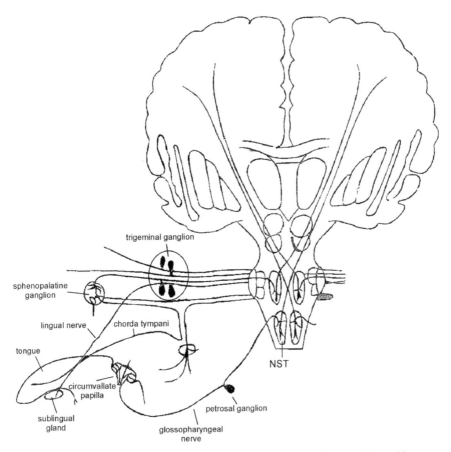

FIGURE 1.6 Early schematic diagram of the peripheral and central taste pathway. Numerous errors are apparent in the schematic, in particular, the pathway of the chorda tympani, which travels with the trigeminal nerve via the sphenopalatine ganglion. In addition, the chorda tympani appears to innervate the sublingual gland and not the taste buds on the anterior tongue. Compare this schematic with Figure 1.7. (From Bekhterev, V. M., *Die Leitungsbahnen im Gehirn und Rückenmark,* Georgi, Leipzig 1899, p. 184.)

The greater superficial petrosal nerve was also thought to serve as motor innervation to muscles of the pharynx and soft palate (see, for example, the pathway of the chorda tympani and greater superficial petrosal nerve diagrammed in Figure 1.6, taken from Bekhterev[9]). Dixon[19] provides a detailed analysis of this confusing literature, contributed not only by eminent neuroanatomists but by physiologists and clinicians as well. Both Dixon and later Sheldon[20] clarified the role of the chorda tympani in transmitting gustatory information and provided the pathway diagram of the peripheral taste pathways in use today

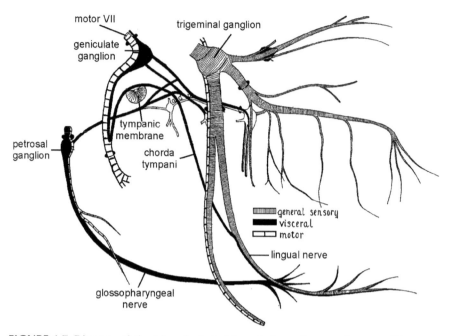

FIGURE 1.7 Diagram of the trigeminal, facial, and glossopharyngeal nerves. The motor components are diagrammed as hash marks. The pathway of the chorda tympani with cell bodies in the geniculate ganglion, travelling through the middle ear (tympanic membrane) and then joining the lingual nerve to innervate the anterior tongue, is now correctly diagrammed. The glossopharyngeal nerve and the greater superficial petrosal branch of the facial nerve innervating palatal taste buds are also correctly diagrammed. (From Sheldon, R. E., *Anat. Rec.,* 3, 617, 1909. With permission.)

(Figure 1.7). However, the role of the chorda tympani in gustatory transmission from the anterior tongue continued to be debated and was further investigated by Schwartz and Weddell in 1938.[18] These investigators once again confirmed the earlier results that the chorda tympani and not the trigeminal nerve innervates taste receptors on the anterior tongue.

By 1900, investigators had detailed the contribution of the facial nerve to the ST via the sensory root or the nervous intermedius of Wrisberg or more simply the nervous intermedius.[11,13,21,22] These investigators concluded that the nervous intermedius of the VIIth nerve contributed to the most rostral extent of the ST. Van Gehuchten[16] examined the relationship of the sensory nerves to the tract by sectioning the Vth and VIIIth nerves central to their ganglia. Finding no evidence of degeneration in the ST, he concluded that these two nerves did not contribute to the tract. However, when the VIIth, IXth, or Xth nerves were sectioned proximal to their ganglia, degeneration occurred in the ST. Van Gehuchten summarized the results of his study in a diagram that succinctly presents the composition of the ST that is similar to the results of more recent investigations (Figure 1.8). The functional role of the nervous intermedius was confirmed by Cushing,[23] who reported on a series of clinical cases in which the trigeminal ganglion had been removed with no resulting loss of taste function.

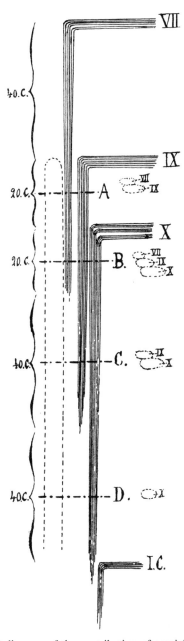

FIGURE 1.8 Longitudinal diagram of the contribution of cranial nerves VII, IX, and X to form the solitary tract. The midline is to the left, and rostral is towards the top of the figure. On the left, the numbers (e.g., 40.C — coups or cuts) indicate the number of 50-μm-thick sections that were cut. The ST is outlined by the dotted line, and representative cross sections indicating the contributions of the cranial nerves are diagrammed as (A–D). This diagram is essentially similar to the results of modern studies using various retrograde tracing techniques. (From Van Gehuchten, A., *Le Névraxe.*, 1, 194, 1900.)

1.4   EARLY DESCRIPTIONS OF THE NUCLEUS
OF THE SOLITARY TRACT

Although the ST had been defined, the surrounding nucleus remained relatively unstudied. Van Gehuchten[24] referred to the nuclear mass associated with the ST as the nucleus tractus solitarius. However, in early anatomical depictions, the nuclei associated with the VIIth, IXth, and Xth nerves were considered to be separated and not a longitudinal column surrounding the ST.

Despite the close anatomical relationship between the ST and the surrounding nucleus, a number of investigators also suggested that the entering sensory afferent fibers do not all terminate in the NST. For example, Gerebtzoff,[25] while describing the terminations of fibers of the ST in the surrounding nucleus of the solitary tract, also stated that the IXth and Xth nerves terminate in the dorsal lateral nucleus of the glossopharyngeal and vagus nerves, as well as in the posterior portion of the triangular nucleus and in the nucleus intercalatus of Staderini. In fact, some authors designated the nucleus intercalatus of Staderini, a nuclear mass situated between the dorsal motor nucleus of the vagus and the hypoglossal nucleus, the gustatory nucleus. For example, Ariëns Kappers, based on his extensive comparative neuroanatomical studies, considered the nucleus intercalatus of Staderini to be a gustatory center in mammals.[26]

This conclusion derives from the finding that the nucleus intercalatus attains greatest development in mammals where taste is "exquisite" and, therefore, indicative of participation in gustatory function. Thus, rodents with many taste buds have a large nucleus intercalatus of Staderini, whereas the almost taste-free dolphin has a small nucleus intercalatus of Staderini. The role of the nucleus intercalatus of Staderini as the gustatory nucleus was also institutionalized in textbooks of neuroanatomy (see, for example, Figure 1.9, where the gustatory projection passes through the solitary nucleus to terminate in a nucleus labeled GUST).[27] More recent neuroanatomical and neurophysiological studies have determined that the nucleus intercalatus of Staderini is part of a brainstem complex organized to integrate inputs related to neck and ocular motor control.[28]

Investigators in the 1940s and 1950s repeated the earlier anatomical investigations, sometimes using newer methodology, or examined the NST in other species. For example, Åström[29] and Torvik[30] studied the central terminations of the trigeminal, facial, glossopharyngeal, and vagus nerves and their nuclei using the Golgi technique in mouse and rat (Figure 1.10). Kerr[31] and Rhoton et al.[32] used the newly developed Nauta-Gygax technique to demonstrate degenerating fibers in the cat and monkey, respectively. Essentially, these later investigations confirmed the earlier studies while adding details on additional species. All of them confirmed the projection pattern of the afferent taste fibers to form the descending ST and the role of the NST in gustatory processing.

The rapid development of neural tracing techniques using the property of nerves to transport various tracers (radioactive amino acids, horseradish peroxidase, lectins, toxins, and fluorescent dyes) was applied to further reexamine the

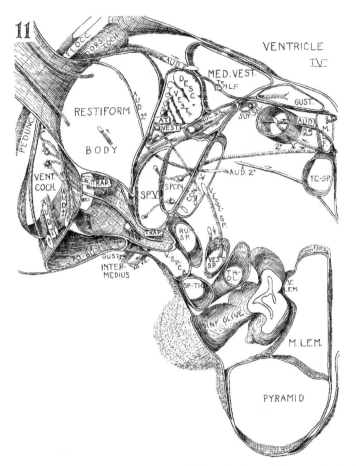

FIGURE 1.9 Diagram of the left side of the medulla at the level of the entering facial nerve. In this diagram, the sensory branch of the facial nerve — the nervus intermedius — passes through the solitary nucleus (NUC. SOL.) to terminate in the "gustatory nucleus (GUST)" situated close to the IVth ventricle. (From Krieg, W. J. S., *Functional Neuroanatomy*, Blakiston, Philadelphia, 1942, Plate 11.)

projection patterns of the gustatory afferent fibers to the NST. Judicious use of these labels has not only confirmed the early neuroanatomical descriptions but has also revealed in amazing precision the fine details of the termination patterns and the extent of the terminal fields of the gustatory afferent fibers (detailed by King in Chapter 2). Both retrograde and anterogradely transported labels applied to cut nerves and ganglia, and injected into receptive fields have all been applied to the peripheral gustatory system. The cumulative results of these recent investigations have confirmed the basic rostral-caudal projection pattern established all early in the 1900s but has also permitted tracing the axons of individual nerves. For example, taste buds on the tongue and palate project to concurrent

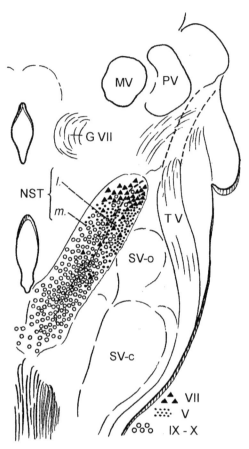

FIGURE 1.10 Diagram of a horizontal section of the right nucleus of the solitary tract (NST, lateral *l* and medial *m*) showing the terminal pattern of the facial (filled triangles, VII), trigeminal (small dots, V), glossopharyngeal, and vagus nerves (open circles, IX–X). G VII, genu of facial nerve; MV, motor trigeminal; PV, principal trigeminal nucleus; SV-c, nucleus caudalis of the spinal trigeminal nucleus; SV-o, nucleus oralis of the spinal trigeminal nucleus; TV, trigeminal tract. (Adapted from Torvik, A., *J. Comp. Neurol.,* 106, 91, 1956. With permission.)

terminal fields in the rNST rather than in separate locations, suggesting functional integration of their sensory information (Figure 1.11).[33] Injection of anterograde tracers into peripheral receptive fields serves as an additional confirmation of the central projection pattern and has the additional advantage that minimal surgery is required to apply the tracers.[34,35] The use of these techniques also permits investigation of the development and influence of environmental factors on the gustatory central projection patterns (see Hill and May in Chapter 6).

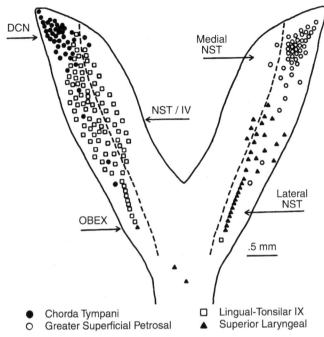

● Chorda Tympani   □ Lingual-Tonsilar IX
○ Greater Superficial Petrosal   ▲ Superior Laryngeal

FIGURE 1.11 Schematic of a horizontal section through the rat medulla to show the termination pattern of the gustatory afferent fibers in the NST. The data in this figure were obtained by anterograde transport of horseradish peroxidase applied to the chorda tympani (VIIth nerve, filled circles), greater superficial petrosal nerve (VIIth nerve, open circles), lingual tonsilar nerve (IXth nerve, open squares), and the superior laryngeal nerve (Xth nerve, filled triangles). Although the termination patterns for the facial, glossopharyngeal, and vagus nerves are bilateral, they have been separated to clearly illustrate the extent of overlap. (From Hamilton, R. B. and Norgren, R. *J. Comp. Neurol.*, 222, 568, 1984. With permission.)

1.5 FIRST ELECTROPHYSIOLOGICAL INVESTIGATIONS OF THE ROSTRAL NUCLEUS OF THE SOLITARY TRACT

Although neuroanatomical tracing studies strongly suggest that the rNST is the site of termination of afferent sensory fibers innervating taste receptors in the oral cavity, confirmation of the role of the rNST in gustatory processing required the application of neurophysiological techniques. This was accomplished in Carl Pfaffmann's laboratory in the late 1950s and appeared in published form in 1961.[36] At the time of these pioneering experiments, Pfaffmann's students had to decide whether to attempt to record from the rNST, which the bulk of the anatomical evidence indicated was the site of origin of the second-order gustatory neurons, or probe the nucleus intercalatus of Staderini, also referred to as the "gustatory nucleus."[27] They chose to record from the rNST and were able to make both multi- and single-unit recordings in rats from neurons that responded to stimulation of the tongue with taste stimuli

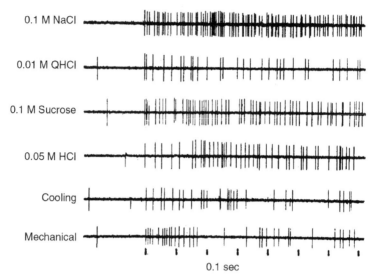

FIGURE 1.12 Extracellular recording from an rNST neuron responding to taste, thermal, and mechanical stimuli applied to the rat tongue. (From Pfaffmann, C. et al., Gustatory discharges in the rat medulla and thalamus, in *Sensory Communication*, Rosenblith, W. A. (Ed.), MIT Press, Cambridge, MA, 1961, p. 462. With permission.)

(Figure 1.12). In a subsequent electrophysiological experiment, the medulla was mapped in rats, and gustatory activity was shown to be limited to the rNST.[37] Electrode penetrations lateral and medial to the rNST did not respond to gustatory stimulation of the oral cavity and surrounding areas. These electrophysiological experiments were the forerunner of many subsequent investigations detailed in this book.

1.6 SUMMARY

The current understanding of the neuroanatomy of the rNST resulted from a number of discoveries spanning 100 years. In the 1850s, early investigators of the medulla first described the ST, which was suggested to have a number of functions, both sensory and motor. Subsequent investigators discovered that the ST was composed of the central processes of the glossopharyngeal and vagus nerves, followed some time later by the finding that the sensory portion of the facial nerve also contributed to the ST. Investigators then began to detail the grey matter surrounding the ST, which at first was thought to consist of separate nuclei, but later was determined to be a continuous column of second-order neurons. Because the afferent input to the rNST originated principally from taste receptors, the rNST was defined as the medullary taste relay nucleus, although other medullary sites were proposed based on comparative neuroanatomy to be the gustatory nucleus. The role of the rNST as the first central nervous system taste relay was cemented by electrophysiological recordings from neurons in the rNST, which were found to respond when taste stimuli were flowed over the tongue.

ACKNOWLEDGMENT

The preparation of this chapter was supported in part by NIH grant DC 000288 from the National Institute on Deafness and Other Communication Disorders to R. M. Bradley.

REFERENCES

1. Patton, H. D. Physiology of smell and taste, *Annu. Rev. Physiol.*, 12, 848, 1950.
2. Stilling, B. *Ueber die Textur der Medulla oblongata.* F. Enke, Erlangen, 1843.
3. Clarke, J. L. Researches on the intimate structure of the brain, human and comparative. First series. On the structure of the medulla oblongata, *Philos. Trans. R. Soc. Lond.*, 148, 231, 1858.
4. Clarke, J. L. Researches on the intimate structure of the brain. Second series, *Philos. Trans. R. Soc. Lond.*, 158, 263, 1868.
5. Meynert, T. The brain of mammals, in *A Manual of Histology,* Stricker, S. (Ed.), Wood, New York, 1872.
6. Duval, M. Recherches sur l'origine réelle des nerfs craniens, *J. Anat. Physiol.* 285, 1880.
7. Krause, C. F. T. *Handbuch der Menschlichen Anatomie.* Hahn, Hannover, 1876.
8. Roller, C. F. W. Der centrale Verlauf des nervus glossopharyngeus-der nucleus lateralis medius, *Arch. Mikroskop. Anat.*, 19, 347, 1881.
9. Bekhterev, V. M. *Die Leitungsbahnen im Gehirn und Rückenmark.* Georgi, Leipzig, 1899.
10. Dees, O. Zur Anatomie und Physiologie des Nervus vagus, in *Archiv. fur Psychiatrie,* 1889, p. 89.
11. Koelliker, A. *Handbuch der Gewebelehr des Menschen.* Engelmann, Leipzig, 1896.
12. Cramer, A. *Beiträge zur feineren Anatomie der medulla oblongata und der Brücke.* Fischer, Jena, 1894.
13. Hiss, W. Die Entwicklung der ersten Nervenbahnen beim menschlichen Embryo, *Arch. Anat. Physiol. Wissensch. Med.* 368, 1887.
14. Turner, W. A. The central connections and relations of the trigeminal, vago-glossopharyngeal, vago-accessory, and hypoglossal nerves, *J. Anat. Physiol.*, 29, 1, 1895.
15. Ramón Y Cajal, S. Ursprünge des Vagus und Glossopharyngeus, in *Beitrag zum Studium der Medulla Oblongata des Kleinhirns und des Ursprungs der Gehirnnerven,* Barth, Leipzig, 1896.
16. Van Gehuchten, A. Recherches sur la termination centrale des nerfs sensibles périphériques, *Le Névraxe*, 1, 173, 1900.
17. Bruce, A. On the dorsal or so-called sensory nucleus of the glossopharyngeal nerve, and on the nuclei of origin of the trigeminal nerve, *Brain*, 21, 383, 1898.
18. Schwartz, H. G. and Weddell, G. Observations on the pathways transmitting the sensation of taste, *Brain*, 61, 99, 1938.
19. Dixon, A. F. The sensory distribution of the facial nerve in man, *J. Anat. Physiol.*, 33, 471, 1898.
20. Sheldon, R. E. The phylogeny of the facial nerve and chorda tympani, *Anat. Rec.*, 3, 593, 1909.
21. Duval, M. Recherches sur l'origine réelle des nerfs craniens. Glosso-pharyngien et nerf de Wrisberg, *J. Anat. Physiol.* 535, 1880.
22. Van Gehuchten, A. Recherches sur la terminaison centrale des nerfs sensibles périphériques. I le nerf intermédiaire de Wrisberg, *Le Névraxe*, 1, 5, 1900.

23. Cushing, H. The taste fibers and their independence of the nervus trigeminus, *Johns Hopkins Hosp. Bull.*, 14, 71, 1903.

24. Van Gehuchten, A. *Anatomie du système nerveux de l'homme.* Uystpruyst-Dieudonné, Louvain, 1897.

25. Gerebtzoff, M. A. Les voies centrales de la sensibilité du goût et leurs terminations thalamiques, *Cellule*, 48, 91, 1939.

26. Ariëns Kappers, C. U., Huber, G. C. and Crosby, E. C. The medulla oblongata, in *The Comparative Anatomy of the Nervous System of Verebrates, Including Man,* Ariëns Kappers, C. U. et al. (Eds.), MacMillan, New York, 1936.

27. Krieg, W. J. S. *Functional Neuroanatomy.* Blakiston, Philadelphia, 1942.

28. Stechison, M. T. and Saint-Cyr, J. A. Organization of spinal inputs to the perihypoglossal complex of the cat, *J. Comp. Neurol.*, 246, 555, 1986.

29. Åström, K. E. On the central course of afferent fibers in the trigeminal, facial, glossopharyngeal, and vagal nerves and their nuclei in the mouse, *Acta Physiol. Scand.*, 29, 209, 1953.

30. Torvik, A. Afferent connections to the sensory trigeminal nuclei, the nucleus of the solitary tract and adjacent structures — an experimental study in the rat, *J. Comp. Neurol.*, 106, 51, 1956.

31. Kerr, F. W. L. Facial, vagal and glossopharyngeal nerves in the cat. Afferent connections, *Arch. Neurol.*, 6, 264, 1962.

32. Rhoton, A. L. Jr., O'Leary, J. L. and Ferguson, J. P. The trigeminal, facial, vagal, and glossopharyngeal nerves in the monkey. Afferent connections, *Arch. Neurol.*, 14, 530, 1966.

33. Hamilton, R. B. and Norgren, R. Central projections of gustatory nerves in the rat, *J. Comp. Neurol.*, 222, 560, 1984.

34. Altschuler, S. M., Bao, X., Bieger, D., Hopkins, D. A. and Miselis, R. R. Viscerotopic representation of the upper alimentary tract in the rat: sensory ganglia and nuclei of the solitary and spinal trigeminal tracts, *J. Comp. Neurol.*, 283, 248, 1989.

35. Bradley, R. M., Mistretta, C. M., Bates, C. A. and Killackey, H. P. Transganglionic transport of HRP from the circumvallate papilla of the rat, *Brain Res.*, 361, 154, 1985.

36. Pfaffmann, C., Erickson, R. P., Frommer, G. P. and Halpern, B. P. Gustatory discharges in the rat medulla and thalamus, in *Sensory Communication,* Rosenblith, W. A. (Ed.), MIT Press, Cambridge, MA, 1961.

37. Makous, W., Nord, S., Oakley, B. and Pfaffmann, C. The gustatory relay in the medulla, in *Olfaction and Taste,* Zotterman, Y. (Ed.), Pergamon Press, Oxford, U.K., 1963.

38. Bianchi, A. L., Denavit-Saubie, M. and Champagnat, J. Central control of breathing in mammals: neuronal circuitry, membrane properties, and neurotransmitters, *Physiol. Rev.*, 75, 1, 1995.

2 Anatomy of the Rostral Nucleus of the Solitary Tract

Michael S. King

CONTENTS

2.1 INTRODUCTION

The nucleus of the solitary tract (NST) is a major sensory nucleus in the dorsal medulla that receives cardiovascular, visceral, respiratory, gustatory, and orotactile information.[1,2] The NST begins at the level of the pyramidal decussation near the cervical spinal cord and extends rostral to the caudal part of the dorsal cochlear nucleus.[3] Caudally, the NST spans the midline forming the commissural subnucleus. Rostral to the area postrema, the NST splits into left and right halves that straddle the midline and abut the ventrolateral edge of the fourth ventricle. The traditional anatomical beginning of the rostral NST (rNST) is the point where the medial edge of the nucleus no longer contacts the fourth ventricle (Figure 2.1). The rNST initially is displaced laterally from the fourth ventricle by the medial vestibular and prepositus hypoglossal nuclei and at its rostral extreme is adjacent to the spinal trigeminal and vestibular nuclei. Although histologically contiguous with the caudal part of the

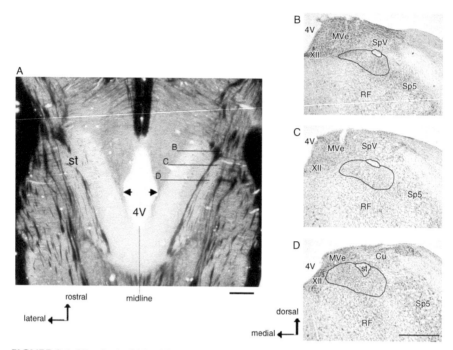

FIGURE 2.1 Histological identification of the rNST. (A) The medulla in a horizontal tissue section stained for myelin (provided by Drs. D. L. Hill and C. T. King). The arrowheads indicate the traditional beginning of the rNST. (B–D) The rNST in Nissl-stained coronal tissue sections at the three rostral-caudal levels indicated in A (provided by Dr. C. T. King). The scale bars indicate 1 mm and abbreviations are 4V, fourth ventricle; Cu, cuneate nucleus; MVe, medial vestibular nucleus; RF, reticular formation; Sp5, spinal trigeminal nucleus; SpV, spinal vestibular nucleus; st, solitary tract; XII, hypoglossal nucleus.

nucleus, the rNST is functionally distinct because it receives gustatory and other orosensory primary afferent input.[4,5] In fact, the traditional anatomical location of the border between the rostral and caudal NST only partly reflects the functional differences between these regions of the nucleus. Because terminal fields of orosensory cranial nerves extend several hundred microns caudal to where the NST moves lateral to the fourth ventricle,[6–10] functionally the rNST extends caudal to its traditional beginning.

2.2 HISTOLOGY OF THE rNST

2.2.1 HISTOLOGICAL IDENTIFICATION OF THE rNST

In the horizontal plane, the NST appears V-shaped, with the commissural portion of the nucleus forming the base of the V and the rNST the apices (Figure 2.1A).[3] In this plane, the solitary tract is clearly visible running within the lateral part of the nucleus. In coronal sections, the rNST is situated just lateral and ventral to the medial and spinal vestibular nuclei, just medial to the spinal trigeminal nucleus and dorsal to the parvocellular reticular formation (Figures 2.1B–D).

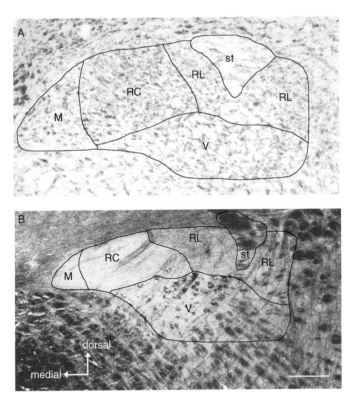

FIGURE 2.2 Illustration of rNST subdivisions in coronal tissue sections stained for Nissl (A) and myelin (B). The scale bar is 250 μm. Abbreviations are M, medial subdivision; RC, rostral central subdivision; RL, rostral lateral subdivision; st, solitary tract; and V, ventral subdivision. (A is from tissue provided by Dr. C. T. King, and the image in B was provided by Dr. S. P. Travers.)

Histologically, the rNST is distinct from the surrounding region due to the relatively small size of neuronal somata within this nucleus.[11–13] This distinction is particularly clear along the medial and dorsal boundaries of the rNST. However, there are some large neurons, particularly within the ventral half of the rNST, which are similar in size to neurons in the subjacent reticular formation, making the delineation of the ventral border of this nucleus difficult (Figure 2.2A). In addition, the lateral border is hard to define in Nissl-stained tissue. The solitary tract is near the dorsolateral border of the rNST, but there is a small portion of the nucleus lateral to the tract. The lateral border is most distinct in tissue stained for myelinated fibers because of the abundance of fibers just lateral to the rNST (Figure 2.2B).[12,13]

Even though the ventral region of the rNST contains fascicles running in the rostral-caudal direction, Weil-stained preparations help define the ventral border of this nucleus because the fascicles within the reticular formation are much more regularly arranged.[13] Several other staining techniques have been used in an attempt to clearly delineate the borders of the rNST, although with limited success. For example, stains for mitochondrial enzymes like NADH dehydrogenase, cytochrome oxidase, and succinate dehydrogenase preferentially label neuropil within the terminal field of the chorda tympani nerve,[10,14,15] whereas immunohistochemical labeling for neurotransmitters, neurotrophins, and their receptors[16–30] tends to label populations of neurons or neuropil scattered throughout the nucleus. Some of these techniques make part of the border of the rNST clear, but none delineate the whole perimeter of the nucleus. However, results from studies using various anatomical techniques indicate that the rNST is a histologically diverse nucleus and define possible subdivisions within it.

2.2.2 IDENTIFICATION AND FUNCTION OF rNST SUBDIVISIONS

Neurons within the rNST are not homogenous nor are they arranged in an orderly manner. However, histological, cytoarchitectural, and connectivity differences among neurons in different regions of the rNST have led to the separation of the nucleus into subdivisions. Initially, based on differences in the size of neuronal somata and cell density as revealed in Nissl-stained tissue, the rat rNST was split into medial and lateral subdivisions.[7] More recent studies, using a combination of anatomical techniques including Nissl-staining, reduced silver staining to examine neuropil, the Golgi method to reveal neuron morphology, as well as tract tracing to determine nuclear inputs and outputs, have designated four or five subdivisions in the rat or hamster rNST, respectively (Figure 2.2).[10,12,13]

Although these subdivisions are differentiated on anatomical bases, they also are functionally distinct. Therefore, the separation of the rNST into subdivisions may be important for the elucidation of the mechanisms of sensory processing within this nucleus. However, because in most studies a combination of anatomical techniques is not used, clear delineation of the subdivision boundaries may not be possible. So, recently some researchers have resorted to dividing the rNST into six subfields based solely on the dimensions of the rNST.[31,32] These subfields are simply a convenient and reliable way to subdivide the rNST to facilitate discussion of functional differences among populations of neurons within the nucleus, but they do not reflect any specific anatomical features. Therefore, because the subdivisions, which are based on anatomical variation within the rNST, seem to have functional significance, it is important to consider them here.

The medial (M) subdivision is the smallest, occupying the medial-most quarter of the rNST, just lateral to the medial vestibular nucleus.[12,13] M contains small, moderately packed neurons with almost no neuropil. Although some preganglionic parasympathetic neurons are scattered within the ventromedial portion of M,[33] most are just outside this subdivision and must be considered separate from the rNST due to their motor function. Neurons in M do not receive direct cranial nerve input, but some are interconnected with the caudal NST,[34,35] possibly linking gustatory and viscerosensory information.

The rostral central (RC) subdivision is immediately lateral to M, medial to the solitary tract, and occupies the middle third to half of the dorsal part of the rNST.[12,13] It contains small- to medium-sized neurons that are densely packed and tend to cluster. Neuropil staining in this subdivision is very light. There are a few myelinated fibers that pass through RC from the overlying vestibular nuclei, particularly in the lateral parts of this subdivision. RC receives the densest afferent input from cranial nerves[10,12] and contains the majority of neurons that contribute to ascending gustatory pathways.[13, 27,36]

The rostral lateral (RL) subdivision surrounds the solitary tract and extends to the lateral edge of the rNST.[12,13] It contains neurons similar in size to those in RC except that they are more loosely packed and do not cluster as much. There is a significant increase in the number of myelinated fibers traversing this subdivision in the dorso-ventral direction as compared to RC. Neurons in RL receive afferent fibers from cranial nerves;[10,12] the majority of these fibers transmit oral tactile information.[37] RL also contains some neurons that project to the parabrachial nucleus (PBN) in the pons.[13,27,36]

The ventral (V) subdivision is located ventral to both RC and RL and extends to the subjacent reticular formation.[12,13] This subdivision is distinct due to large, sparsely packed neurons that are loosely arranged around many rostral-caudal running fiber bundles that traverse this part of the nucleus. V contains the majority of rNST neurons that project to brainstem oromotor centers[13,34,38] as well as some neurons that project rostrally to the PBN.[13,27,36]

A dorsal (D) subdivision has been described in the hamster.[12] D is a thin strip just dorsal to RC and ventral to the vestibular nuclei. The neurons in D are small and very sparse, and the terminal fields of cranial nerves avoid this part of the rNST.[10,12]

The caudal NST also is split into several subdivisions.[2,12,39, 40] Some of the caudal NST subdivisions are specific to that region of the nucleus, whereas others are contiguous with rNST subdivisions. For example, there is no rostral counterpart to the dorsolateral, laminar, and ventrolateral subdivisions, which are prominent in the caudal NST.[12] Of the rNST subdivisions, M and V are viewed as continuous longitudinal columns of cells with similar cytoarchitecture throughout the whole nucleus, whereas RC and RL are anatomically distinct from the caudal central (CC) and caudal lateral (CL) subdivisions.[12] Neurons within CC and CL tend to be larger and have different morphology and orientation as compared to neurons in RC and RL. Although there are anatomical distinctions, there is no clear line between CC and RC or CL and RL. Instead, the changes in neuronal size, shape, and orientation occur gradually, beginning just caudal to where the NST separates from the fourth ventricle. This region of gradual change is an area of the caudal NST that receives input from the lingual branch of the IX nerve,[9,12] so it is interesting to speculate that the anatomical changes are related to the gradual change in input received by neurons in this region from cranial nerves.

2.2.3 NEURON TYPES WITHIN THE rNST

In an attempt to understand the structural organization of the rNST, the morphology of neurons in this nucleus has been visualized using a variety of techniques.[11–13,16,41–44] Based on cell soma size and shape, and the number of primary dendrites and dendritic tree morphology apparent in Golgi impregnated tissue, it has been proposed that there are three main neuron types within the rNST: ovoid, multipolar, and elongate (Figure 2.3). Injections of horseradish peroxidase into taste-responsive regions of

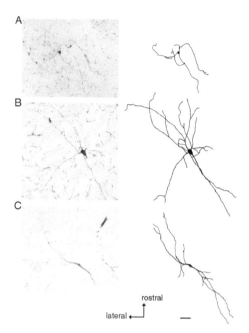

rostral

lateral

FIGURE 2.3 A biocytin-filled ovoid (A), multipolar (B), and elongate (C) rNST neuron and accompanying reconstructions. These cells were labeled with biocytin during whole-cell recording in living rat brain slices. The scale bar is 100 µm. (From King, M. S. and Bradley, R. M., *J Comp Neurol.*, 344, 50, 1994. Used with permission.)

the rNST, which can cause Golgi-like filling of neurons, also led to the identification of these three types of cells.[45] Labeling of individual neurons after electrophysiological evaluation has been accomplished by adding biocytin to the recording electrode.[46,47] The latter technique is particularly powerful because the whole dendritic tree of a single neuron is labeled and can be analyzed and related to its biophysical properties.

Retrograde labeling of rNST neurons following tracer injections into target nuclei has the advantage of identifying projection neurons within the rNST but typically leads to the visualization of only small neuronal regions, so in these studies neuronal classification depends mainly on somata shape and proximal dendrite morphology.[13,27,44] Data from studies using these various techniques are consistent with the three-group classification scheme for rNST neurons described below. However, the functional significance of each of the morphological groups has not been clearly defined, suggesting that there may be better methods of categorizing rNST neurons.

The smallest cells have spherical or ovoid somata and typically two or three thin primary dendrites (0.75 µm thick on average). The dendrites of these ovoid cells usually are relatively short, extending 180 µm from the soma on average,[15,46] and therefore stay within a localized region of the nucleus. Due to this morphology as well as evidence that many ovoid neurons are GABAergic and do not project out of the rNST,[13,16,19,36] these cells are considered to be interneurons. Each rNST subdivision contains numerous ovoid neurons, implying a significant role for these cells in the processing of information throughout the nucleus.

Multipolar (or stellate) neurons have relatively large, pyramidal- or polygonal-shaped somata with three to five primary dendrites. The dendrites can extend long distances (average dendritic length of 650 μm)[46] within the rNST from one subdivision to the next, as well as sometimes out of the nucleus, most typically from V into RC or the reticular formation.[12] Although emanating from the cell body in all directions, the dendrites of some multipolar neurons are predominantly oriented in the horizontal plane and can extend several hundred microns in the rostral-caudal direction.[42] This dendritic orientation places these cells perpendicular to afferent fibers exiting the solitary tract and implies that they receive convergent input, a characteristic frequently observed in rNST neurons in electrophysiological studies.[48–50] Multipolar neurons are abundant at taste-responsive sites in RC,[45] and although relatively spine-free, their dendrites have more spines than the other cell types in this nucleus,[46,51] suggesting that multipolar neurons receive primary afferent input. In fact, an analysis at the electron microscopic level has indicated that multipolar neurons in RC receive synapses that resemble primary fiber endings.[52] Retrograde labeling studies have identified multipolar neurons as projection neurons in the rNST, providing most of the ascending and descending output from the nucleus.[13,36]

Elongate (or fusiform) neurons have a distinctive cell body that is larger in one dimension than the other, with two thick primary dendrites that project from opposite poles of the cell (1.1 μm thick on average).[12,13,46] Typically, the dendrites are long (average dendritic length of 800 μm), relatively unbranched, and spineless with a predominant medial-lateral orientation.[12,46] Elongate neurons are present at taste-responsive sites in RC and RL[45] and receive primary-like synaptic endings.[52] Similar to multipolar neurons, elongate cells are retrogradely labeled following tracer injections into the parabrachial nucleus and the reticular formation.[13,36] One of the distinguishing characteristics of RL is the abundance of elongate neurons,[12] suggesting a role for this cell type in the processing of nongustatory orosensory input.

Due to variability in labeling techniques and probable preferential labeling of larger neurons, the relative number of ovoid, multipolar, and elongate neurons in the rNST can only be estimated. However, it appears that ovoid cells are most abundant, making up just under 50% of the neurons.[11,46] Although the distribution of multipolar and elongate neurons within the rNST is not homogenous,[12] overall the former outnumber the latter, with less than 20% of the neurons being elongate.

Although ovoid, multipolar, and elongate neurons are morphologically distinct, they may not represent functionally distinct groups of cells in the rNST. In fact, unlike these cell types in the caudal NST,[53] ovoid, multipolar, and elongate rNST neurons do not appear to have distinct biophysical properties or responses to neurotransmitters.[46] However, it is clear that there is one functional distinction among these neuron types in the rNST: The smaller ovoid neurons are interneurons, whereas the larger multipolar and elongate neurons are projection neurons. Therefore, neuronal size is possibly the best indicator of function. It is unclear whether there are other functionally meaningful, morphologically identifiable groups of neurons in the rNST. It is possible that each of the three broad categories of cell types consists of subgroups of functionally distinct neurons that may be differentiable on the basis of soma size,[54] dendritic complexity,[52] or nuclear profile.[43] In addition, it also is possible that other classification schemes would more effectively discern functional groups of rNST neurons. For example, rNST neurons separated into morphologically distinct groups based on total cell volume, soma area, mean dendritic segment length, dendritic swelling density, spine density, and number of primary dendrites have a few physiological differences related to the breadth of responses to gustatory stimulation.[47,55,56]

2.3 CONNECTIONS OF THE rNST

The initial investigations of inputs from the oral cavity to the brainstem identified degenerating fibers and synaptic boutons after cranial nerve lesions in the mouse[6] and rat.[7] A similar technique initially was used to trace the projections from the rNST to the pons and thalamus in rat.[57] The details of these projections have been determined by several more recent studies employing a variety of tract-tracing techniques, including autoradiography of transported tritiated amino acids[38,58–60] and visualization of transported horseradish peroxidase,[8,10,16,34,61–64] fluorescent dyes,[13,44,65,66] and viral proteins[67–71] in rat, hamster, rabbit, cat, lamb, and monkey. Labeling with multiple fluorescent dyes in the same tissue has been particularly useful in identifying subpopulations of rNST neurons that project to different targets[13] as well as to simultaneously label terminal fields of different nerves,[66] whereas viral labeling techniques are useful in identifying higher order afferents to a particular target due to trans-synaptic transport.[67–71] The details of the inputs and outputs of the rNST are described in this section.

2.3.1 INPUTS OF THE rNST

Sensory information from the oral cavity is transmitted to the rNST by the facial, glossopharyngeal, vagus, and trigeminal nerves.[6–9,59–64,72–75] Fibers from these nerves enter the solitary tract and most travel caudally, giving off branches that terminate within the NST (Figure 2.4). The terminal fields of these cranial nerves are very dense and form an overlapping topographic pattern within the rNST.[9] The chorda tympani and greater superficial petrosal branches of the facial nerve inner-vate the anterior tongue and palate, respectively, and project to the most rostral portions of the nucleus. The terminals of these branches of the facial nerve inter-mingle within a region of the dorsal rNST just medial to the solitary tract, an area corresponding to RC and medial RL.[12] The lingual-tonsilar branch of the glossopha-ryngeal nerve transmits sensory information from the posterior tongue to a region of the rNST that overlaps the caudal part of the facial nerve terminal field and extends into the caudal NST. Similar to facial nerve input, neurons in RC and the medial part of RL receive the vast majority of the orosensory input from the glossopharyngeal nerve.

Although most of the terminal field of the superior laryngeal branch of the vagus nerve, which innervates the epiglottis, is within the caudal NST, there are a small number of fibers from this nerve near the glossopharyngeal terminal field in the

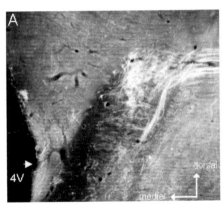
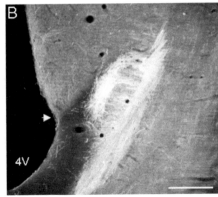

FIGURE 2.4 Horizontal sections showing the terminal fields of the chorda tympani branch of the facial nerve (A) and the lingual-tonsilar branch of the glossopharyngeal nerve (B) in the rat rNST. The terminal fields were labeled with horseradish peroxidase and are viewed in dark field. The fourth ventricle (4V) and traditional beginning of the rNST (arrowhead) are indicated. The scale bar is 1 mm. (Images provided by Drs. D. L. Hill and C. T. King.)

caudal-most part of the rNST.[9] Finally, the lingual branch of the trigeminal nerve transmits tactile information from the oral epithelium to the rNST. The trigeminal terminal field avoids the rostral extreme of the rNST, being concentrated just lateral to the glossopharyngeal nerve input, primarily within RL.[8,10,12,60]

Electrophysiological studies complement and extend the anatomical data. For example, neurons that respond to stimulation of the anterior tongue are located within the rostral half of the rNST, whereas responses to posterior tongue stimulation are more abundant in the caudal half of the rNST.[37,76] In addition, neurons that preferentially respond to oral tactile stimulation are located more posterior and lateral to neurons that prefer gustatory stimuli.[76–78] Finally, indicative of the overlap of termination sites of different gustatory nerves within the rat rNST, about half of the taste-responsive neurons in the rNST respond to stimulation of more than one receptor region.[37,48]

Because the solitary tract travels in the rostral-caudal direction within the dorsolateral portion of the rNST and the main input to rNST neurons via cranial nerves branch into the nucleus from the solitary tract, the orientation of the dendritic trees of rNST neurons may be functionally significant. Although most data points to a preferential orientation of dendritic trees of rNST neurons in the horizontal plane,[12,42,46] most neurons have significant dendritic extensions in the coronal plane as well. Although the main dendritic orientation may be biased by the plane of section in which the neurons were evaluated, most rNST neurons do seem to extend farthest in the rostral-caudal direction, along the long axis of the nucleus. This orientation positions the dendritic tree perpendicular to its main input from cranial nerves and may promote integration of input from different cranial nerves.

Neurons within the rNST receive input from several brain regions (Figure 2.5). Within the brainstem, the main sources of input are the caudal NST[35] and the medial parabrachial nucleus.[79–81] The projection from the caudal NST originates in general

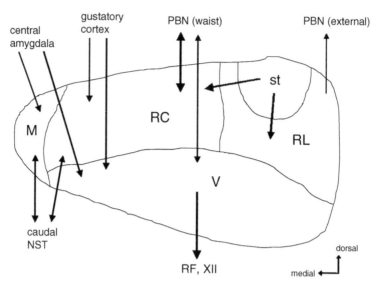

FIGURE 2.5 Schematic illustration of inputs and outputs of the rNST. The thickness of the arrow indicates the relative density of the projection (for example, the projection from RC to the waist area of the PBN is heavier than the projection from V). Only direct projections are shown, and areas in the ventral forebrain and hypothalamus are not included. Abbreviations are as in Figures 2.1 and 2.2.

viscerosensory regions of the nucleus and terminates primarily within RC and M.[35] This intranuclear connection may allow the interaction of visceral and gustatory information within the rNST.[79] The projection from the parabrachial nucleus in the pons arises from neurons within the "waist" region,[81] the primary pontine taste-responsive area,[82,83] and terminates primarily within RC and V.[81] These anatomical data suggest that input to the rNST from the pons could alter the processing of gustatory information within this nucleus as well as its ascending and descending output.

Forebrain projections to the rNST originate in areas responsible for higher-order sensory processing and behavioral or hormonal responses. The insular cortex, particularly the rostral gustatory region, sends fibers bilaterally to the rNST.[35,84,85] These fibers preferentially terminate in RC and V[35] and, therefore, are presumed to influence ascending and descending output of the nucleus. Electrophysiological analyses of the effects of projections from the gustatory cortex on responses of rNST neurons to taste stimulation suggest that the cortical influence can alter the processing of sensory input within the rNST as well as the transmission of information out of the nucleus.[86,87] Fibers originating in the central nucleus of the amygdala terminate throughout the rNST, but are particularly dense within M and V, suggesting that this descending input could preferentially regulate rNST output to brainstem oromotor centers.[88] Afferent projections to the rNST from the bed nucleus of the stria terminalis, paraventricular nucleus, lateral hypothalamus, and prefrontal cortex have been demonstrated anatomically.[35,89] The details of these connections with the rNST and their influence on neurons in the nucleus are currently being explored.[90–92]

2.3.2 OUTPUTS OF THE rNST

As the first central structure to receive orosensory input, one of the main roles of the rNST is to disseminate orosensory information to brain areas involved in sensory perception and the generation of appropriate behavioral and physiological responses. Accordingly, neural pathways originating in the rNST ascend to higher brain centers and descend to medullary oromotor nuclei (Figure 2.5; for review, see Reference 5). In rodents and lagomorphs, the ascending pathway includes an obligatory synapse in the caudal parabrachial nucleus in the pons.[4,36,38,58,79,81,93] This projection arises from multipolar and elongate rNST neurons mainly located within RC.[13,27,36,42] About 60% of the neurons in RC can be retrogradely labeled following injection of a neural tracer into the caudal PBN.[36] Implying that a major role of the rNST is to convey gustatory information to the PBN, estimates from electrophysiological data are that between 31% and 80% of taste-responsive rNST neurons project to the PBN.[76,90,94] The rNST-PBN projection is primarily ipsilateral[36,38,58] with a small contralateral component[65] and terminates within the waist area and external region of the PBN.[79,93] Most taste-responsive neurons in the PBN are located within the waist area, which includes the central medial and ventral lateral subnuclei.[82,83,95,96] Neurons in the external PBN primarily respond to gustatory or tactile stimulation of the posterior oral cavity.[96] From the PBN, the majority of the ascending orosensory pathway proceeds to the ventral posteromedial thalamus[95,97–99] and then to the insular cortex.[100,101] However, a minor projection from the waist area of the PBN proceeds directly to the insula,[94,101–103] whereas other projections terminate in the lateral hypothalamus, substantia inominata, bed nucleus of the stria terminalis, and central nucleus of the amygdala.[98,104,105]

The major medullary projection from the rNST is to the parvocellular reticular formation,[34,38,106,107] an area that contains neurons that project to oromotor brainstem nuclei.[67,107–110] Within the rNST, multipolar and elongate neurons in V are the main source of this projection.[12,13] Other descending projections terminate within the caudal NST, in salivatory nuclei, and directly in brainstem oromotor nuclei. The projection to the caudal NST originates primarily in RC and M[13] and ends in intermediate regions of the nucleus that receive glossopharyngeal and vagal input.[38,67] This intranuclear connection may provide a mechanism for orosensory input to influence visceral sensation and oromotor output. Although there may be sparse direct projections from the rNST to several brainstem oromotor nuclei,[34,58,111] the only significant direct pathway is to the hypoglossal nucleus.[38,71,107] Projections from the rNST to the motor trigeminal and facial nuclei and the nucleus ambiguus are primarily indirect.[68–70] In conclusion, through direct and indirect pathways, rNST neurons can influence medullary motor output, thereby producing the appropriate oromotor response to gustatory and oral tactile input.

2.4 FUNCTIONAL NEUROANATOMY OF THE rNST

2.4.1 FOS-IMMUNOHISTOCHEMICAL STUDIES OF THE rNST

As summarized above, data from neural tract tracing studies indicate that neurons in RC and the medial part of RL receive afferent input from cranial nerves that

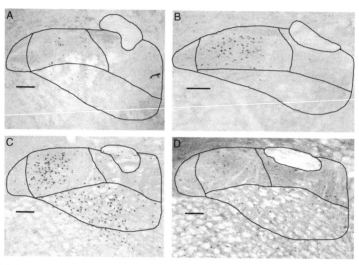

FIGURE 2.6 Images showing Fos-immunoreactive neurons within rNST subdivisions in coronal sections following electrical stimulation of the chorda tympani branch of the facial nerve (A) and the lingual-tonsilar branch of the glossopharyngeal nerve (B) in anesthetized rats. The section shown in A is from just caudal to the dorsal cochlear nucleus, and the section in B is about 300 μm caudal to that. The subdivisions are not labeled but are outlined as in Figure 2.2. Notice that the labeled neurons are mainly confined to RC. Images show Fos-IR neurons within the rNST induced by intraoral infusion of 3 mM quinine (C) and distilled water (D) in conscious rats. There are many labeled neurons in medial RC and V after stimulation with quinine. The scale bars indicate 100 μm. (Images in A and B provided by Dr. T. A. Harrison; tissue for the images in C and D provided by Drs. C. T. King and A. C. Spector.)

innervate the oral cavity. Electrophysiological data confirm these findings and help to define populations of rNST neurons that respond to specific oral stimuli.[37,112,113] Immunohistochemical detection of the Fos protein, the product of the immediate early gene *c-Fos*,[114,115] has been used to identify rNST neurons activated by cranial nerve and gustatory stimulation.[31,32,116–119] Although the identification of Fos-immunoreactive (Fos-IR) neurons is not always clear, and has other technical limitations,[120] it is useful for identifying possible functionally distinct neuronal populations. The finding that electrical stimulation of the chorda tympani and the lingual nerve in anesthetized rats leads to Fos-IR neurons within their terminal fields, mainly within RC (Figures 2.6A and 2.6B),[119] demonstrates the usefulness of this technique within the rNST.

Importantly, the localization of Fos-IR neurons has determined that different tastants activate different populations of neurons within RC (Figures 2.6C and 2.6D). These cell populations are not completely distinct, but suggest a rough chemotopic distribution of rNST neurons.[32] Specifically, the intraoral infusion of the bitter-tasting quinine elicits Fos-IR neurons concentrated in the medial portion of RC, whereas neurons activated by sucrose (sweet) or citric acid (sour) are more evenly distributed within this subdivision.[31,32,116,121] Because quinine elicits significantly more gapes (stereotypical large mouth openings in response to aversive compounds[122]) than the other tastants, the data suggest that neurons in the medial part of RC may be necessary for the stimulation of gaping behavior.[32,116] Supporting this idea, transection of the glossopharyngeal nerves, which eliminates gaping behavior, also decreases the number of Fos-IR neurons in the medial RC.[31,123] These results imply that the neurons activated in the rNST by tastants are related to the resulting behavioral response. However, the pattern of Fos-IR neurons following oral stimulation with NaCl (salty) is not significantly different from the pattern elicited by water[32] even though there are distinct electrophysiological and behavioral responses to this tastant.[112,113,122,124]

As oromotor responses to the oral infusion of tastants are ultimately caused by the activation of medullary motor centers,[110,125] a necessary step in defining the neural circuit underlying these behaviors is to link rNST neurons that respond to specific tastants to motor neurons. As discussed above, most rNST projections to motor neurons are indirect, synapsing within the caudal NST or reticular formation first. Furthermore, most rNST neurons that project to the reticular formation are located within V, separate from the neurons in RC that are activated directly by taste input. Putative axon terminals of neurons in V have been shown to intermingle with neurons in the reticular formation that are activated by oral infusion of quinine.[121] Therefore, the sensory neurons in medial RC that preferentially respond to oral infusion of quinine may project to V to link to the motor output apparatus. In fact, results of a recent study using the Fos-IR technique suggest that most of the taste-responsive neurons in rNST have intranuclear projections.[126] In the future, it will be necessary to elucidate the details of the connections between sensory neurons and motor circuits. It is predicted that rNST neurons in RC that preferentially respond to different tastants will have different outputs that underlie the disparate behavioral responses that the tastants elicit. Experiments using Fos immunohistochemistry also have addressed the alterations in activated NST neurons following taste aversion learning[127–129] and stimulation of other brain regions[130,131] as well as the neurotransmitter content of neurons activated by different tastants.[132,133]

2.4.2 Synapse Morphology and Distribution within the rNST

Studies of synapse morphology and distribution within the rNST suggest that the processing of orosensory information within this nucleus is quite complex and heterogeneous among nuclear regions. In particular, the arrangement of primary synapses from the chorda tympani nerve in glomerulus-like endings indicates that the effects of afferent input would be modulated on the dendrites of second-order neurons in the rostral regions of the rNST (Figure 2.7).[25,134] These synaptic complexes typically include a dendritic spine with a single primary ending and two presynaptic terminals from other sources. Dopaminergic neurons in RC may be preferred targets of these inputs.[25] Occasionally, presynaptic endings contact primary synaptic terminals,[134] implying that presynaptic modulation occurs, as well. Indicating that both excitatory and inhibitory mechanisms shape the responses of rNST

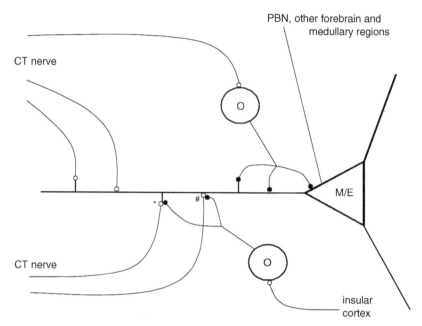

FIGURE 2.7 Schematic illustration of the distribution and organization of synapses on a multipolar or elongate (M/E) neuron within the chorda tympani (CT) terminal field in the rNST. Thin lines represent axons and thick lines, dendrites. Based on work by Davis[25] and Whitehead,[52,134] presumed excitatory synaptic endings (open circles) preferentially terminate on distal dendritic processes, whereas inhibitory endings (closed circles), primarily from local ovoid neurons (O), tend to contact proximal dendrites and somata. Input from the insular cortex may influence inhibitory interneurons[86] and presumed inhibitory endings also may contact presynaptic terminals (#). In addition, some synaptic complexes include a postsynaptic spine, primary fiber ending, and presumed inhibitory terminal arranged in a glomerulus-like complex (*).

neurons within the chorda tympani terminal field, asymmetric primary endings and symmetrical synapses are typically part of the glomerulus-like complex.

On the other hand, the most common synaptic relationship within the terminal field of the glossopharyngeal nerve is between a primary ending and a single dendritic spine.[135] These data suggest that the processing of sensory input from the anterior and posterior oral cavity within the rNST is different, with the former relying on more complex synaptic interactions involving both excitatory and inhibitory processes. Synaptic responses in neurons receiving glossopharyngeal input may not be subject to strong inhibitory influence and, therefore, may cause more rapid, consistent behavioral responses, most likely to bitter compounds.

It is estimated that about 45% of the synapses within the rNST contain the inhibitory neurotransmitter GABA.[136,137] This high number of GABAergic synapses is consistent with the prevalence of the inhibitory influence of GABA on rNST neurons.[138,139] Presumed GABAergic synapses preferentially contact the soma or proximal dendrites of PBN projection neurons within RC.[52] This synaptic distribution is in contrast to the position of excitatory primary-like endings on spines of distal dendrites and is optimal for inhibitory modulation of excitatory input to these neurons (Figure 2.7). Clearly, the synaptology indicates that the ascending output of the rNST is the result of complex processing within second-order neurons. Unfortunately, nothing is known about synapses that contain other neurotransmitters within the rNST.

2.5 SUMMARY

The traditional border between the rostral and caudal portions of the NST, where the nucleus moves lateral to the fourth ventricle, is actually a transition zone where the input to the nucleus changes from primarily viscerosensory to orosensory. Rostral to this point, the rNST is histologically, cytoarchitecturally, and functionally distinct. Neuronal somata within the rNST tend to be smaller than those in surrounding structures as well as the caudal NST, and the rNST is relatively devoid of axon bundles except for the solitary tract located along the dorsolateral border, rostral-caudal running fascicles in the ventral part of the nucleus, and fibers from the overlying vestibular nuclei that traverse the lateral half of the rNST. Based on neuronal size, morphology, and distribution, the rat rNST has been split into four subdivisions. Although there is considerable overlap of neuron types and connectivity of these subdivisions, there are some functional distinctions. For example, the rostral central subdivision (RC) receives most of the gustatory input from the periphery and contains the majority of neurons that project to the pons, whereas most of the projections to medullary oromotor centers originate in the ventral subnucleus. Three main neuronal types have been identified within the rNST: ovoid, multipolar, and elongate. These neuron types are distributed throughout the nucleus and seem to serve different roles, with ovoid neurons as inhibitory interneurons and multipolar and elongate neurons as projection neurons. The identification of other morphologically distinct subpopulations of neurons with defined functions would improve the understanding of the central processing of orosensory information.

Orosensory input enters the rNST in a topographic fashion from the trigeminal, facial, glossopharyngeal, and vagus nerves. Although separate at some points within the nucleus, these inputs converge significantly upon rNST neurons and extend caudal to the traditional beginning of the rNST. RC contains most of the neurons responsive to gustatory input, whereas tactile orosensory input preferentially terminates within the rostral lateral subdivision. The main ascending gustatory pathway that leads to perception and other higher-order functions proceeds from the rNST to the parabrachial nucleus in the pons, the ventral posteromedial thalamus, and then to the insular cortex. Other ascending pathways originating in the rNST terminate in the ventral forebrain, lateral hypothalamus, and amygdala. Projections from the rNST also terminate within the medullary reticular formation where oromotor behaviors are organized.

Immunohistochemical localization of the Fos protein has been used to assign functions to particular groups of neurons within the rNST. Direct stimulation of gustatory cranial nerves and intraoral infusions of tastants elicit distinct patterns of Fos-immunoreactive neurons within the rNST, suggesting that different ensembles of neurons within this nucleus subserve different functions. The complex synaptic arrangements within the chorda tympani terminal field as well as the distribution of different synapse types along the dendrites of rNST neurons suggest that the processing of orosensory information within the rNST is very complex. Determining the populations of rNST neurons activated by specific orosensations, and how they process this input, is interconnected; coordinating the physiological and behavioral responses is a challenge of future research.

REFERENCES

1. Loewy, A. D., Central autonomic pathways, in *Central Regulation of Autonomic Functions*, Loewy, A. D. and Speyer, K. M., (Eds.), Oxford University Press, Oxford, 1990, Chap. 6.
2. Jean, A., Le noyau du faisceau solitaire: aspects neuroanatomiques, neurochimiquies et functionnels, *Arch. Int. Physiol. Biochim. Biophys.*, 99, A3, 1991.
3. Paxinos, G. and Watson, G., *The Rat Brain in Stereotaxic Coordinates*, 2nd ed., Academic Press, New York, 1986.
4. Norgren, R., Central neural mechanisms of taste, in *Handbook of Physiology—The Nervous System*, Darien-Smith, I. (Ed.), *Section III*, Brookhart, J. and Mountcastle, V. (Eds.), American Physiological Society, Washington, D.C., 1984, p. 1087.
5. Norgren, R., Gustatory system, in *The Rat Nervous System*, Paxinos, G. (Ed.), Academic Press, New York, 1995, Chap. 29.
6. Åstrom, K. E., On the central course of afferent fibers in the trigeminal, facial, glossopharyngeal, and vagal nerves and their nuclei in the mouse, *Acta Physiol. Scand. Suppl.*, 106, 209, 1953.
7. Torvik, A., Afferent connections of the sensory trigeminal nuclei, the nucleus of the solitary tract and adjacent structures: an experimental study in the rat, *J. Comp. Neurol.*, 106, 51, 1956.
8. Whitehead, M. C. and Frank, M. E., Anatomy of the gustatory system in the hamster: central projections of the chorda tympani and the lingual nerve, *J. Comp. Neurol.*, 220, 378, 1983.
9. Hamilton, R. B. and Norgren, R., Central projections of gustatory nerves in the rat, *J. Comp. Neurol.*, 222, 560, 1984.
10. Barry, M. A., Halsell, C. B., and Whitehead, M. C., Organization of the nucleus of the solitary tract in the hamster: acetylcholinesterase, NADH dehydrogenase, and cytochrome oxidase histochemistry, *Microsc. Res. Tech.*, 26, 231, 1993.
11. Davis, B. J. and Jang, T., The gustatory zone of the nucleus of the solitary tract in the hamster: light microscopic morphometric studies, *Chem. Senses*, 11, 213, 1986.
12. Whitehead, M. C., Neuronal architecture of the nucleus of the solitary tract in the hamster, *J. Comp. Neurol.*, 276, 547, 1988.
13. Halsell, C. B., Travers, S. P., and Travers, J. B., Ascending and descending projections from the rostral nucleus of the solitary tract originate from separate neuronal populations, *Neuroscience*, 72, 185, 1996.
14. Lasiter, P. S. and Kachele, D. L., Elevated NADH-dehydrogenase activity characterizes the rostral gustatory zone of the solitary nucleus in rat, *Brain Res. Bullet.*, 22, 777, 1989.
15. Lasiter, P. S., Wong, D. M., and Kachele, D. L., Postnatal development of the rostral solitary nucleus in rat: dendritic morphology and mitochondrial enzyme activity, *Brain Res. Bullet.*, 22, 313, 1989.
16. Lasiter, P. S. and Kachele, D. L., Organization of GABA and GABA-transaminase containing neurons in the gustatory zone of the nucleus of the solitary tract, *Brain Res. Bull.*, 21, 623, 1988.

17. Hironaka, T. et al., Localization of $GABA_A$-receptor α_1 subunit mRNA-containing neurons in the lower brainstem of the rat, *Molec. Brain Res.*, 7, 335, 1990.
18. Merlio, J. P. et al., Molecular cloning of rat trkC and distribution of cells expressing messenger RNAs for members of the trk family in the rat central nervous system, *Neuroscience*, 51, 513, 1992.
19. Davis, B. J., GABA-like immunoreactivity in the gustatory zone of the nucleus of the solitary tract in the hamster: light- and electron-microscopic studies, *Brain Res. Bull.*, 30, 69, 1993.
20. Davis, B. J. and Kream, R. M., Distribution of tachykinin- and opioid-expressing neurons in the hamster solitary nucleus: an immuno- and *in situ* hybridization histochemical study, *Brain Res.*, 616, 6, 1993.
21. Sweazey, R. D., Distribution of aspartate and glutamate in the nucleus of the solitary tract of the lamb, *Exp. Brain Res.*, 105, 241, 1995.
22. Sweazey, R. D., Distribution of GABA and glycine in the lamb nucleus of the solitary tract, *Brain Res.*, 737, 275, 1996.
23. Yan, Q. et al., Immunocytochemical localization of TrkB in the central nervous system of the adult rat, *J. Comp. Neurol.*, 378, 135, 1997.
24. Yan, Q., Rosenfeld, R. D., Matheson, C. R., Hawkins, N., Lopez, O. T., Bennett, L., and Welcher, A. A., Expression of brain-derived neurotrophic factor protein in the adult rat central nervous system, *Neuroscience*, 78, 431, 1997.
25. Davis, B. J., Synaptic relationships between the chorda tympani and tyrosine hydroxy-lase-immunoreactive dendritic processes in the gustatory zone of the nucleus of the solitary tract in the hamster, *J. Comp. Neurol.*, 392, 78, 1998.
26. Davis, B. J. and Smith, H. M., Neurokinin-1 receptor immunoreactivity in the nucleus of the solitary tract in the hamster, *NeuroReport*, 10, 1003, 1999.
27. Gill, C. F. et al., A subpopulation of neurons in the rat rostral nucleus of the solitary tract that project to the parabrachial nucleus express glutamate-like immunoreactivity, *Brain Res.*, 821, 251, 1999.
28. Heck, W. L., Renehan, W. E., and Schweitzer, L., Redistribution and increased specificity of $GABA_B$ receptors during development of the rostral nucleus of the solitary tract, *Int. J. Devl. Neurosci.*, 19, 503, 2001.
29. King, M. S., Distribution of immunoreactive GABA and glutamate receptors in the gustatory portion of the nucleus of the solitary tract in rat, *Brain Res. Bull.*, 20, 241, 2003.
30. Harrison, T. A., Hoover, D. B., and King, M. S., Distinct regional distributions of NK1 and NK3 neurokinin receptor immunoreactivity in rat brainstem gustatory centers, *Brain Res. Bull.*, 63, 7, 2004.
31. King, C. T. et al., Glossopharyngeal nerve transsection eliminates quinine stimulated Fos-like immunoreactivity in the nucleus of the solitary tract: implications for a functional topography of gustatory nerve input in rats, *J. Neurosci.*, 19, 3107, 1999.
32. Travers, S. P., Quinine and citric acid elicit distinctive Fos-like immunoreactivity in the rat nucleus of the solitary tract, *Amer. J. Physiol.*, 282, R1798, 2002.
33. Contreras, R. J., Gomez, M. M., and Norgren, R., Central origin of cranial nerve para-sympathetic neurons in the rat, *J. Comp. Neurol.*, 190, 373, 1980.
34. Beckman, M. E. and Whitehead, M. C., Intramedullary connections of the rostral nucleus of the solitary tract in hamster, *Brain Res.*, 557, 265, 1991.
35. Whitehead, M. C., Bergula, A., and Holliday, K., Forebrain projections to the rostral nucleus of the solitary tract in the hamster, *J. Comp. Neurol.*, 422, 429, 2000.
36. Whitehead, M. C., Subdivisions and neuron types of the nucleus of the solitary tract that project to the parabrachial nucleus in the hamster, *J. Comp. Neurol.*, 301, 554, 1990.

37. Travers, S. P. and Norgren, R., Organization of orosensory responses in the nucleus of the solitary tract of the rat, *J. Neurophysiol.*, 73, 2144, 1995.

38. Travers, J. B., Efferent projections from the anterior nucleus of the solitary tract of the hamster, *Brain Res.*, 457, 1, 1988.

39. Kalia, M. and Sullivan, J. M., Brainstem projections of sensory and motor components of the vagus nerve in the rat, *J. Comp. Neurol.*, 211, 248, 1982.

40. Loewy, A. D. and Burton, H., Nuclei of the solitary tract: efferent projections to the lower brain stem and spinal cord of the cat, *J. Comp. Neurol.*, 181, 421, 1978.

41. Bradley, R. M., Neurobiology of the gustatory zone of the nucleus tractus solitarius, in *Nucleus of the Solitary Tract*, Barraco, I.R.A. (Ed.), CRC Press, Boca Raton, FL, 1994, Chap. 3.

42. Davis, B. J., Computer-generated rotation analyses reveal a key three-dimensional feature of the nucleus of the solitary tract, *Brain Res. Bull.*, 20, 545, 1988.

43. Davis, B. J. and Jang, T., Tyrosine hydroxylase-like and dopamine β-hydroxylase-like immunoreactivity in the gustatory zone of the nucleus of the solitary tract in the hamster: light- and electron-microscopic studies, *Neuroscience*, 27, 949, 1988.

44. Lasiter, P. S., Effects of early postnatal receptor damage on dendritic development in gustatory recipient zones of the rostral nucleus of the solitary tract, *Dev. Brain Res.*, 61, 197, 1991.

45. Whitehead, M. C. et al., Morphological types of neurons located at taste-responsive sites in the solitary nucleus of the hamster, *Microsc. Res. Tech.*, 26, 245, 1993.

46. King, M. S. and Bradley, R. M., Relationship between structure and function of neurons in the rat rostral nucleus tractus solitarii, *J. Comp. Neurol.*, 344, 50, 1994.

47. Renehan, W. E. et al., Structure and function of gustatory neurons in the nucleus of the solitary tract. I. A classification of neurons based on morphological features, *J. Comp. Neurol.*, 347, 531, 1994.

48. Travers, S. P., Pfaffmann, C., and Norgren, R., Convergence of lingual and palatal gustatory neural activity in the nucleus of the solitary tract, *Brain Res.*, 365, 305, 1986.

49. Sweazey, R. D. and Smith, D. V., Convergence onto hamster medullary taste neurons, *Brain Res.*, 408, 173, 1987.

50. Sweazey, R. D. and Bradley, R. M., Responses of neurons in the lamb nucleus tractus solitarius to stimulation of the caudal oral cavity and epiglottis with different stimulus modalities, *Brain Res.*, 480, 133, 1989.

51. Mistretta, C. M. and Labyak, S. E., Maturation of neuron types in nucleus of solitary tract associated with functional convergence during development of taste circuits, *J. Comp. Neurol.*, 345, 359, 1994.

52. Whitehead, M. C., Distribution of synapses on identified cell types in a gustatory subdivision of the nucleus of the solitary tract, *J. Comp. Neurol.*, 332, 326, 1993.

53. Dekin, M. S., Getting, P. A., and Johnson, S. M., *In vitro* characterization of neurons in the ventral part of the nucleus tractus solitarius. I. Identification of neuronal types and repetitive firing properties, *J. Neurophysiol.*, 58, 195, 1987.

54. King, C. T. and Hill, D. L., Neuroanatomical alterations in the rat nucleus of the solitary tract following early maternal NaCl deprivation and subsequent NaCl repletion, *J. Comp. Neurol.*, 333, 531, 1993.

55. Schweitzer, L. et al., Cell types in the rostral nucleus of the solitary tract, *Brain Res. Rev.*, 20, 185, 1995.

56. Renehan, W. E. et al., Structure and function of gustatory neurons in the nucleus of the solitary tract. II. Relationships between neuronal morphology and physiology, *J. Comp. Neurol.*, 367, 205, 1996.

57. Norgren, R. and Leonard, C. M., Ascending central gustatory pathways, *J. Comp. Neurol.*, 150, 217, 1973.

58. Norgren, R., Projections from the nucleus of the solitary tract in the rat, *Neuroscience*, 3, 207, 1978.

59. Beckstead, R. M and Norgren, R., An autoradiographic examination of the central distribution of the trigeminal, facial, glossopharyngeal, and vagal nerves in the monkey, *J. Comp. Neurol.*, 184, 455, 1979.

60. Contreras, R., Beckstead, R., and Norgren, R., An autoradiographic examination of the central distribution of the trigeminal, facial, glossopharyngeal and vagus nerves, *J. Auton. Nerv. Syst.*, 6, 303, 1982.

61. Nomura, S. and Mizuno, N., Central distribution of afferent and efferent components of the chorda tympani in the cat as revealed by the horseradish peroxidase method, *Brain Res.*, 214, 229, 1981.

62. Bradley, R. M. et al., Transganglionic transport of HRP from the circumvallate papilla of the rat, *Brain Res.*, 361, 154, 1985.

63. Sweazey, R. D. and Bradley, R. M., Central connections of the lingual-tonsillar branch of the glossopharyngeal nerve and the superior laryngeal nerve in lamb, *J. Comp. Neurol.*, 245, 471, 1986.

64. Hanamori, T. and Smith, D. V., Gustatory innervation in the rabbit: central distribution of sensory and motor components of the chorda tympani, glossopharyngeal, and superior laryngeal nerves, *J. Comp. Neurol.*, 282, 1, 1989.

65. Williams, J. B. et al., Demonstration of a bilateral projection from the rostral nucleus of the solitary tract to the medial parabrachial nucleus in the rat, *Brain Res.*, 737, 231, 1996.

66. Mangold, J. E. and Hill, D. L., Postnatal development of gustatory nerve terminal fields in control rats, *Chem. Senses* (Abstract), 30, 467, 2005.

67. Travers, J. B., Montgomery, N., and Sheridan, J., Transneuronal labeling in hamster brainstem following lingual injections with herpes simplex virus-1, *Neuroscience*, 68, 1277, 1995.

68. Fay, R. A. and Norgren, R., Identification of rat brainstem multisynaptic connections to the oral motor nuclei using pseudorabies virus. I. Masticatory muscle motor systems, *Brain Res. Rev.*, 25, 255, 1997.

69. Fay, R. A. and Norgren, R., Identification of rat brainstem multisynaptic connections to the oral motor nuclei using pseudorabies virus. II. Facial muscle motor systems, *Brain Res. Rev.*, 25, 276, 1997.

70. Fay, R. A. and Norgren, R., Identification of rat brainstem multisynaptic connections to the oral motor nuclei using pseudorabies virus. III. Lingual muscle motor systems, *Brain Res. Rev.*, 25, 291, 1997.

71. Travers, J. B. and Rinaman, L., Identification of lingual motor control circuits using two strains of pseudorabies virus, *Neuroscience*, 115, 1139, 2002.

72. Nomura, S. and Mizuno, N., Central distribution of afferent and efferent components of the glossopharyngeal nerve: an HRP study in the cat, *Brain Res.*, 236, 1, 1982.

73. Housley, G. D. et al., Brain stem projections of the glossopharyngeal nerve and its carotid sinus branch in the rat, *Neuroscience*, 22, 237, 1987.

74. Satoda, T. et al., The sites of origin and termination of afferent and efferent components in the lingual and pharyngeal branches of the glossopharyngeal nerve in the Japanese monkey (*Macaca fuscata*), *Neurosci. Res.*, 24, 385, 1996.

75. Hayakawa, T. et al., Subnuclear distribution of afferents from the oral, pharyngeal and laryngeal regions in the nucleus tractus solitarii of the rat: a study using transganglionic transport of cholera toxin, *Neurosci. Res.*, 39, 221, 2001.

76. McPheeters, M. et al., Taste-responsive neurons and their locations in the solitary nucleus of the hamster, *Neuroscience*, 34, 745, 1990.

77. Ogawa, H., Imoto, T., and Hayama, T., Responsiveness of solitaro-parabrachial relay neurons to taste and mechanical stimulations applied to the oral cavity in rats, *Exp. Brain Res.*, 54, 349, 1984.

78. Halsell, C. B., Travers, J. B., and Travers, S. P., Gustatory and tactile stimulation of the posterior tongue activate overlapping but distinctive regions within the nucleus of the solitary tract, *Brain Res.*, 632, 161, 1993.

79. Karimnamazi, H., Travers, S. P., and Travers, J. B., Oral and gastric input to the parabrachial nucleus of the rat, *Brain Res.*, 957, 193, 2002.

80. Krukoff, T. L., Harris, K. H., and Jhamandas, J. H., Efferent projections from the parabrachial nucleus demonstrated with the anterograde tracer *Phaseolus vulgaris* leucoagglutinin, *Brain Res. Bull.*, 30, 163, 1993.

81. Karimnamazi, H. and Travers, J. B., Differential projections from gustatory responsive regions of the parabrachial nucleus to the medulla and forebrain, *Brain Res.*, 813, 283, 1998.

82. Norgren, R. and Pfaffmann, C., The pontine taste area in the rat, *Brain Res.*, 91, 99, 1975.

83. Van Buskirk, R. L. and Smith, D. V., Taste sensitivity of hamster parabrachial nuclei to taste mixtures, *J. Neurophysiol.*, 45, 144, 1981.

84. Shipley, M. T., Insular cortex projection to the nucleus of the solitary tract and brainstem visceromotor regions in the mouse, *Brain Res. Bull.*, 8, 139, 1982.

85. Hayama, T. and Ogawa, H., Two loci of the insular cortex project to the taste zone of the nucleus of the solitary tract in rats, *Neurosci. Letts.*, 303, 49, 2001.

86. DiLorenzo, P. M. and Monroe, S., Corticofugal influence on taste responses in the nucleus of the solitary tract in the rat, *J. Neurophysiol.*, 74, 258, 1995.

87. Smith, D. V. and Li, C.-S., GABA-mediated corticofugal inhibition of taste-responsive neurons in the nucleus of the solitary tract, *Brain Res.*, 858, 408, 2000.

88. Halsell, C. B., Differential distribution of amygdaloid input across rostral solitary nucleus subdivisions in rat, *Ann. N. Y. Acad. Sci.*, 855, 482, 1998.

89. Van der Kooy, D. et al., The organization of projections from the cortex, amygdala, and hypothalamus to the nucleus of the solitary tract in rat, *J. Comp. Neurol.*, 224, 1, 1994.

90. Cho, Y. K., Li, C.-S., and Smith, D. V., Gustatory projections from the nucleus of the solitary tract to the parabrachial nuclei in the hamster, *Chem. Senses*, 27, 81, 2002.

91. Li, C.-S., Cho, Y. K., and Smith, D. V., Taste responses of neurons in the hamster solitary nucleus are modulated by central nucleus of the amygdala, *J. Neurophysiol.*, 88, 2979, 2002.

92. Smith, D. V., Ye, M.-K., and Li, C.-S., Medullary taste responses are modulated by the bed nucleus of the stria terminalis, *Chem. Senses*, 30, 421, 2005.

93. Herbert, H., Moga, M. M., and Saper, C. B., Connections of the parabrachial nucleus with the nucleus of the solitary tract and the medullary reticular formation in the rat, *J. Comp. Neurol.*, 293, 540, 1990.

94. Monroe, S. and DiLorenzo, P. M., Taste responses in neurons in the nucleus of the solitary tract that do and do not project to the parabrachial pons, *J. Neurophysiol.*, 74, 249, 1995.

95. Fulwiler, C. E. and Saper, C. B., Subnuclear organization of the efferent connections of the parabrachial nucleus in the rat, *Brain Res. Rev.*, 7, 229, 1984.

96. Halsell, C. B. and Travers, S. P., Anterior and posterior oral cavity responsive neurons are differentially distributed among parabrachial subnuclei in rat, *J. Neurophysiol.*, 78, 920, 1997.

97. Norgren, R. and Leonard, C. M., Taste pathways in rat brainstem, *Science*, 173, 1136, 1971.

98. Norgren, R., Taste pathways to hypothalamus and amygdala, *J. Comp. Neurol.*, 116, 12, 1976.

99. Bester, H. et al., Differential projections to the intralaminar and gustatory thalamus from the parabrachial area: a PHA-L study in the rat, *J. Comp. Neurol.*, 405, 421, 1990.

100. Wolf, G., Projections of thalamic and cortical gustatory areas in the rat, *J. Comp. Neurol.*, 132, 519, 1968.

101. Norgren, R. and Wolf, G., Projections of thalamic gustatory and lingual areas in the rat, *Brain Res.*, 92, 123, 1975.

102. Lasiter, P. S., Glanzman, D. L, and Mensah, P. A., Direct connectivity between pontine taste areas and gustatory neocortex in rat, *Brain Res.*, 234, 111, 1982.

103. Saper, C. B., Convergence of autonomic and limbic connections in the insular cortex of the rat, *J. Comp. Neurol.*, 210, 163, 1982.

104. Norgren, R., Gustatory afferents to ventral forebrain, *Brain Res.*, 81, 285, 1974.

105. Bernard, J. F., Alden, M., and Besson, J. M., The organization of efferent projections from the pontine parabrachial area to the amygdaloid complex: a *Phaseolus vulgaris* leucoagglutinin (PHA-L) study in the rat, *J. Comp. Neurol.*, 329, 201, 1993.

106. Shammah-Lagnado, S. J., Costa, M.S.M.O., and Ricardo, J. A., Afferent connections of the parvocellular reticular formation: a horseradish peroxidase study in the rat, *Neuroscience*, 50, 403, 1992.

107. Streefland, C. and Jansen, K., Intramedullary projections of the rostral nucleus of the solitary tract in the rat: gustatory influences on autonomic output, *Chem. Senses*, 24, 655, 1999.

108. Travers, J. B. and Norgren, R., Afferent projections to oral motor nuclei in the rat, *J. Comp. Neurol.*, 220, 280, 1983.

109. Ter Horst, G. J. et al., Projections from the rostral parvocellular reticular formation to pontine and medullary nuclei in the rat: involvement in autonomic regulation and orofacial motor control, *Neuroscience*, 40, 735, 1991.

110. Travers, J. B., Dinardo, L. A., and Karimnamazi, H., Motor and premotor mechanisms of licking, *Neurosci. Biobehav. Rev.*, 21, 631, 1997.

111. Zerari-Mailly, F. et al., Trigemino-solitarii-facial pathway in rats, *J. Comp. Neurol.*, 487, 176, 2005.

112. Doetsch, G. S. and Erikson, R. P., Synaptic processing of taste-quality information in the nucleus tractus solitarius of the rat, *J. Neurophysiol.*, 33, 490, 1970.

113. Scott, T. R. and Giza, B. K., Coding channels in the taste system of the rat, *Science*, 249, 1585, 1990.

114. Sagar, S. M. et al., Expression of c-Fos protein in brain: metabolic mapping at the cellular level, *Science*, 240, 1328, 1988.

115. Morgan, J. I. and Curran, T., Stimulus-transcription coupling in neurons: role of cellular immediate-early genes, *Trends Neurosci.*, 12, 459, 1989.

116. Harrer, M. I. and Travers, S. P., Topographic organization of Fos-like immunoreactivity in the rostral nucleus of the solitary tract evoked by gustatory stimulation with sucrose and quinine, *Brain Res.*, 711, 125, 1996.

117. Streefland, C. et al., c-Fos expression in the brainstem after voluntary ingestion of sucrose in the rat, *Neurobiology*, 4, 85, 1996.

118. Yamamoto, T. and Sawa, K., c-Fos-like immunoreactivity in the brainstem following gastric loads of various chemical solutions in rats, *Brain Res.*, 866, 135, 2000.

119. Harrison, T. A., Chorda tympani nerve stimulation evokes Fos expression in regionally limited neuron populations within the gustatory nucleus of the solitary tract, *Brain Res.*, 904, 54, 2001.

120. Sharp, F. R., Sagar, S. M., and Swanson, R. A., Metabolic mapping with cellular resolution: c-Fos vs. 2-deoxyglucose, *Crit. Rev. Neurobiol.*, 7, 205, 1993.

121. DiNardo, L. A. and Travers, J. B., Distribution of Fos-like immunoreactivity in the medullary reticular formation of the rat after gustatory elicited ingestion and rejection behaviors, *J. Neurosci.*, 17, 3826, 1997.

122. Grill, H. J. and Norgren, R., The taste reactivity test: I. Mimetic responses to gustatory stimuli in neurologically normal rats, *Brain Res.*, 143, 263, 1978.

123. King, C. T., Garcea, M., and Spector, A. C., Glossopharyngeal nerve regeneration is essential for the complete recovery of quinine-stimulated oromotor rejection behaviors and central patterns of neuronal activity in the nucleus of the solitary tract in the rat, *J. Neurosci.*, 20, 8426, 2000.

124. Smith, D. V., Travers, J. B., and Van Buskirk, R. L., Brainstem correlates of gustatory similarity in the hamster, *Brain Res. Bull.*, 4, 359, 1979.

125. Travers, J. B., Urbanek, K., and Grill, H. J., Fos-like immunoreactivity in the brain stem following oral quinine stimulation in decerebrate rats, *Am. J. Physiol.*, 277, R384, 1999.

126. Travers, S. P. and Hu, H., Extranuclear projections of rNST neurons expressing gustatory-elicited Fos, *J. Comp. Neurol.*, 427, 124, 2000.

127. Houpt, T. A. et al., Increased c-Fos expression in nucleus of the solitary tract correlated with conditioned taste aversion to sucrose in rats, *Neurosci. Letts.*, 172, 1, 1994.

128. Swank, M. W. and Bernstein, I. L., c-Fos induction in response to a conditioned stimulus after single trial taste aversion learning, *Brain Res.*, 636, 202, 1994.

129. Yamamoto, T. and Sawa, K., Comparison of c-Fos-like immunoreactivity in the brainstem following intraoral and intragastric infusions of chemical solutions in rats, *Brain Res.*, 866, 144, 2000.

130. Petrov, T., Jhamandas, J. H., and Krukoff, T. L., Connectivity between brainstem autonomic structures and expression of c-Fos following electrical stimulation of the central nucleus of the amygdala, *Cell Tissue Res.*, 283, 367, 1996.

131. King, M. S. et al., The location of Fos-immunoreactive neurons following electrical stimulation of the gustatory parabrachial nucleus in conscious rats, *Soc. Neurosci. Abstracts*, 594, 8, 2003.

132. Davis, B. J. and Smith, D. V., Taste-induced Fos expression in dopaminergic neurons in the nucleus of the solitary tract in the hamster, *Chem. Senses*, 24, 597, 1999.

133. Harrison, T. A. and Hoard, J. L., A small number of gustatory NST neurons activated by electrical stimulation of the posterior tongue taste nerve are GABA-ergic interneurons, *Soc. Neurosci. Abstracts*, 288, 1, 2001.

134. Whitehead, M. C., Anatomy of the gustatory system in the hamster: synaptology of facial afferent terminals in the solitary nucleus, *J. Comp. Neurol.*, 244, 72, 1986.

135. Brining, S. K. and Smith, D. V., Distribution and synaptology of glossopharyngeal afferent nerve terminals in the nucleus of the solitary tract of the hamster, *J. Comp. Neurol.*, 365, 556, 1996.

136. Wetherton, B. M. et al., Structure and function of gustatory neurons in the nucleus of the solitary tract. III. Classification of terminals using cluster analysis, *Biotech. Histochem.*, 73, 164, 1998.

137. Leonard, N. L., Renehan, W. E., and Schweitzer, L., Structure and function of gustatory neurons in the nucleus of the solitary tract. IV. The morphology and synaptology of GABA-immunoreactive terminals, *Neuroscience*, 92, 151, 1999.

138. Smith, D. V., Li, C.-S., and Davis, B. J., Excitatory and inhibitory modulation of taste responses in the hamster brainstem, *Ann. N. Y. Acad. Sci.*, 855, 450, 1998.

139. Wang, L. and Bradley, R. M., *In vitro* study of afferent synaptic transmission in the rostral gustatory zone of the rat nucleus of the solitary tract, *Brain Res.*, 702, 188, 1995.

3 Neurotransmitters and Receptors Expressed by rNST Neurons

Robert M. Bradley and Michael S. King

CONTENTS

3.1 INTRODUCTION

Synaptic processing in the NST has been extensively studied using neurophysiological, immunocytochemical, and microinjection techniques. Numerous neurotransmitters and neuromodulators have been shown to be involved in the nongustatory NST,[1–3] but studies of neurotransmitters involved in the rostral NST are more limited.[4–6] Because the NST extends over a considerable rostral-caudal distance and investigators restrict their analyses to a particular function of the NST (e.g., cardiovascular control or taste processing), descriptions are often limited to a region of the NST responsible for that function. Moreover, it is not always possible, when investigators use coronal sections, to determine if the rostral NST was studied (see, for example, Ambalavanar et al., 1998 and Saha et al., 2001)[7,8] even though the title of the report implies that the whole NST is being investigated. When investigators use horizontal sections of the entire extent of the NST, differences in neurotransmitter distribution become evident (see Davis and Jang, 1988).[9] Thus, even though numerous neurotransmitters or neuromodulators have been shown to be involved in NST processing, closer examination of these publications reveals that they are limited primarily to caudal nongustatory NST.

3.2 NEUROTRANSMITTERS IN THE AFFERENT TASTE PATHWAY

3.2.1 IMMUNOCYTOCHEMICAL STUDIES

Afferent input to the NST derived from gustatory receptors begins with synaptic activity in primary afferent fibers with cell bodies in the geniculate, petrosal, and nodose ganglia. Neurotransmitters released from the primary synapses serve to either excite or modulate the activity of NST neurons. Although glutamate has been shown to be the principal transmitter released from the primary afferent endings, other transmitters have been demonstrated in the cell bodies of neurons in the petrosal and geniculate ganglia using immunocytochemical staining. The petrosal ganglion has been shown to contain glutamate, substance P, tyrosine hydroxylase, vasoactive intestinal polypeptide, calcitonin gene-related peptide, galanin, and aspartate.[10–16] However, the petrosal ganglion contains soma of afferent neurons supplying a number of peripheral sensory organs as well as innervating taste receptors on the posterior tongue. No similar immunocytochemical survey has been undertaken for geniculate ganglion neurons that, while innervating anterior tongue taste buds, also supply tactile receptors behind the ear.

3.2.2 NEUROPHARMACOLOGICAL STUDIES

By using retrograde tracing techniques, neurons that innervate taste buds have been identified in the ganglia. Neurons in the geniculate and petrosal ganglia supplying the tongue, fluorescently labeled with either Fluorogold or true blue, were immunoreacted for glutamate receptors[17] or acutely dissociated and analyzed with patch-clamp recording.[18,19] The pharmacological properties of neurons of the nodose ganglion innervating laryngeal taste buds have not been determined.

Based on considerable research, glutamate is considered to be the primary excitatory neurotransmitter released at the central terminations of the facial, glossopharyngeal, and vagus nerves. Glutamate binds to three types of receptor classified according to their sensitivity to different agonists.[20] The first type of glutamate receptor is sensitive to N-methyl-D-aspartate (NMDA), a second type is sensitive to α-amino-3-hydroxy-5-methyl-4-isoxazoleproprionic acid (AMPA), and a third type is sensitive to kainate. The receptor channels are heteromers composed of a number of subunits.[21,22] The NMDA receptors are composed of various combinations of NR1-3 subunits, and the AMPA receptors are formed by GluR1–4 subunits, whereas kainate receptors are combinations of GluR5, 6, and 7 subunits as well as KA1 and 2 groups.

By using antibodies to NMDA and AMPA receptor subunits, Caicedo et al.[17] were able to demonstrate differences in the distribution of immunoreactive neurons in the geniculate ganglion. The majority of the geniculate ganglion neurons were strongly immunoreactive for GluR2/3, GluR4, and NR1. Although most geniculate ganglion neurons retrogradely labeled by injection of tracer into the tongue also were immunoreactive for GluR2/3, GluR4, and NR1, neurons double labeled with tracer and immunoreaction product for GluR1 subunits were much less prevalent (Figure 3.1). It is possible that not all the ganglion neurons innervating the taste buds could be labeled by the use of tongue injections, possibly leading to an underrepresentation of immunolabeled geniculate taste neurons. Specifically, it is

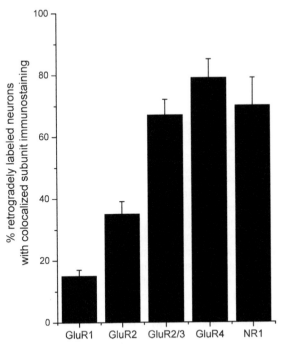

FIGURE 3.1 Percentage of retrogradely labeled neurons in the rat geniculate ganglion that immunostain with antibodies to AMPA and NMDA glutamate receptor subunits. (From Caicedo, A. et al., *Chem. Senses*, 29, 463, 2004. With permission.)

likely that taste buds in the most anterior clefts of the foliate papilla would not be labeled, and therefore ganglion neurons innervating the foliate papilla would potentially be immunoreacted but not retrogradely labeled.

There were significant differences in the response of acutely isolated petrosal and geniculate ganglion neurons innervating the anterior and posterior tongue to the application of acetylcholine (ACh), serotonin (5-HT), substance P (SP), and γ-aminobutyric acid (GABA). Whereas petrosal ganglion neurons responded to ACh, 5-HT, SP, and GABA, geniculate ganglion neurons only responded to SP and GABA (Figure 3.2).[19] Geniculate ganglion neurons retrogradely labeled by injection of

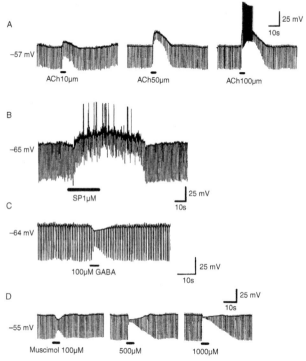

FIGURE 3.2 Voltage responses recorded from petrosal and geniculate ganglion neurons innervating the tongue to hyperpolarizing injections of –100 pA, 100 ms current pulses. (A) Membrane responses of a petrosal ganglion neuron to a 5 sec application (bar) of increasing concentrations of acetylcholine (ACh). The ACh application induced membrane depolarization, accompanied by a decrease in input resistance. The response is dose-dependent and returns to control levels within 30 sec after termination of application. The depolarization resulting from the highest concentration of ACh initiated a burst of action potentials. (B) Membrane response of a petrosal ganglion neuron to a 5 sec application (bar) of substance P (SP). In this neuron, SP application depolarized the neuron and initiated the production of action potentials. (C) Membrane response of a geniculate ganglion neuron to a 5 sec application (bar) of γ-aminobutyric acid (GABA). In this neuron, GABA application hyperpolarized the neuron with a decrease in input resistance. (D) Membrane responses of a petrosal ganglion neuron to a 5 sec application (bar) of increasing concentrations of the GABA$_A$ receptor agonist muscimol. The muscimol application induced dose-dependent membrane hyperpolarization, accompanied by a decrease in input resistance. (From Koga, T. and Bradley, R. M., *J. Neurophysiol.*, 84, 1404, 2000, with permission.)

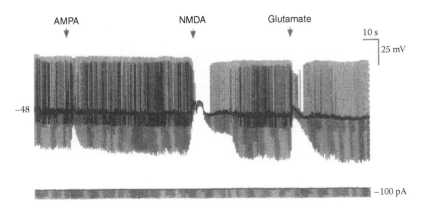

FIGURE 3.3 Voltage responses recorded from a geniculate ganglion neuron innervating the tongue to hyperpolarizing injections of −100 pA, 100 ms current pulses. The positive going potentials are action potentials that follow the hyperpolarizing current pulse (rebound spikes). Application of a 10 ms pulse of AMPA, NMDA, and glutamate all transiently depolarized the neuron, decreased its input resistance, and reduced the number of rebound spikes. NMDA and glutamate have a greater effect on this neuron than AMPA. (Reprinted from King, M. S. and Bradley, R. M., *Brain Res.* 866, 237, 2000, with permission from Elsevier.)

tracer into the anterior tongue and soft palate and by applying tracer to the cut posterior auricular branch innervating the skin behind the pinna were isolated and patch-clamped to determine responses to glutamate.[18] Dissociated geniculate ganglion neurons from all three receptive fields responded to one or more of the NMDA and AMPA glutamate receptor agonists. More neurons respond to NMDA than to AMPA, and a neuronal response to one of the agonists did not necessarily mean that it responded to the other (Figure 3.3).

The results of both the immunocytochemical and electrophysiological studies are in agreement, suggesting that a number of neuropeptide and neurotransmitter receptors are present on the soma of the gustatory ganglion neurons. It is generally assumed that the peripheral and central projections of the ganglion cells also express the same receptors. Bidirectional transport of NMDA receptors from the soma to the central and peripheral branches has been demonstrated in the vagus nerve,[23] and NMDA receptor immunoreactivity has been described in vagal synapses within the NST.[24] In addition, glossopharyngeal nerve axon terminals in the taste buds have been shown to express GluR2/3 receptor subunits.[17] The presynaptic glutamate receptors expressed in central terminals of the gustatory nerves are thought to act as autoreceptors involved in the regulation of transmitter release.[25]

3.3 SYNAPTIC TRANSMISSION IN rNST

3.3.1 IMMUNOCYTOCHEMICAL STUDIES OF GLUTAMATE IN rNST

Although the role of many of these neurotransmitters and neuromodulators present in the geniculate and petrosal ganglion soma remains to be defined, the function of glutamate and GABA has been well studied. Both anatomical and electrophysiological

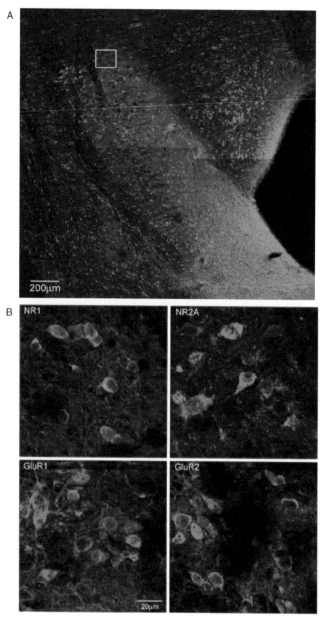

FIGURE 3.4 (See color insert following p.78) (A) Low-power photomicrograph of a horizontal section through the left half of the rat medulla. The 4th ventricle is on the right and rostral is towards the top of the micrograph. The section has been immunoreacted with antibodies to the NR2 glutamate receptor subtype. The NST immunoreacts strongly for the NR2 receptor subtype. The solitary tract is unstained. The small white square indicates the location of the higher power photomicrographs in (B). (B) Immunoreactions to the NR1, NR2A, GluR1, and GluR2 glutamate receptor subtypes. Neurons in the NST express all four of these glutamate receptor subtypes.

techniques have been used to define the neurotransmitters and neuromodulators and their receptors within the rNST. As the major excitatory neurotransmitter in the mammalian brain, it is not surprising that glutamate is present in neurons and puncta throughout the rNST (Figure 3.4).[26,27] Ovoid, multipolar, and elongate neurons are all glutamate-immunoreactive and distributed throughout the subdivisions of the rNST.[26] In addition, unlabelled rNST neurons are often surrounded by glutamate-reactive puncta (Figure 3.5). The distribution of the glutamate receptor subtypes is not uniform within the rNST,[28,29] leading to the suggestion that they may play different functional roles in the processing of orosensory information and its dissemination from the nucleus.

For example, neurons expressing AMPA and NMDA receptors are abundant throughout the rNST, including the RC subdivision (Figure 3.6), whereas neurons expressing kainate receptors are sparsely encountered. These anatomical data are consistent with the physiological findings that AMPA receptors play a particularly important role in responses to cranial nerve input and suggest a prominent role for NMDA receptors as well. About 20% of the neurons in the RC subdivision (see Chapter 2 for definitions of rNST subdivisions) that contribute to ascending taste pathways express either AMPA or NMDA receptors, implying that some rNST neurons that receive glutamatergic afferent input directly send axons to the PBN.[28] Therefore, a minority of rNST neurons may rapidly transmit orosensory input to higher brain centers. Metabotropic glutamate receptors are not as abundant in the rNST as the ionotropic receptors; however, some neurons express the metabotropic 8 receptor subtype, particularly in the V subdivision,[28] suggesting that activation of these receptors could modulate rNST output to the reticular formation.

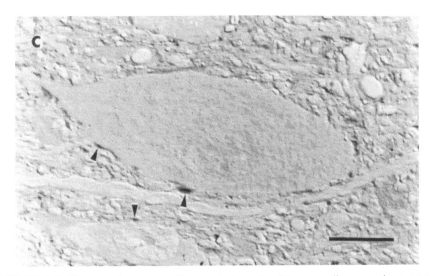

FIGURE 3.5 Glutamate immunoreactive puncta (arrowheads) surrounding a non-immunoreactive neuron in the rNST. (From Sweazey, R. D., *Exp. Brain Res.*, 105, 241, 1995, with kind permission of Springer Science and Business Media.)

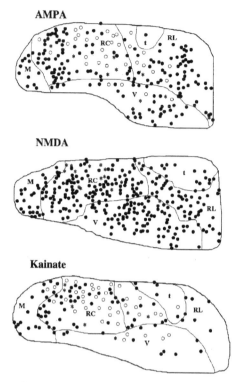

FIGURE 3.6 Diagram of the distribution of AMPA, NMDA, and kainate immunoreactive neurons in the rNST. Open circles represent the location of rNST neurons that project to a parabrachial nucleus, asterisks indicate a parabrachial nucleus projecting neuron that is also immunoreactive for a glutamate receptor, and filled circles are the locations of immunoreactive neurons. (Reprinted from King, M. S., *Brain Res. Bull.*, 60, 241, 2003, with permission from Elsevier.)

3.3.2 NEUROPHYSIOLOGICAL STUDIES OF GLUTAMATE ACTIVITY IN rNST

3.3.2.1 *In Vitro* Studies

Electrophysiological evidence has accumulated, indicating that glutamate is released from primary afferent terminals[30,31] and influences all morphological cell types in the rNST. By using *in vitro* techniques, glutamate agonists and antagonists can be directly applied to defined populations of neurons that can also be labeled for later morphometric analysis. Acutely isolated rNST neurons retain their morphological characteristics, permitting the application of transmitter agonists and antagonists directly to the neurons.[32,33] In horizontal brain slices of the medulla, the NST and ST can clearly be visualized when illuminated from below (Figure 3.7). With this preparation, patch-clamp electrodes can be accurately positioned and the ST electrically stimulated to evoke postsynaptic potentials.

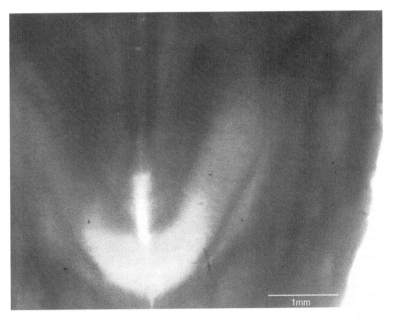

FIGURE 3.7 Photomicrograph of a horizontal brain slice showing the nucleus of the solitary tract transluminated from below. The outline of the NST is clearly visible as well as the darker solitary tract.

These potentials are either a depolarizing excitatory postsynaptic potential (EPSP), a hyperpolarizing inhibitory postsynaptic potential (IPSP), or a mixed EPSP and IPSP response. The postsynaptic potentials are monosynaptic because they are initiated after a short (< 4 ms) and constant latency, have a regular shape with a fast rise time, and follow a 10 Hz stimulation in a 1:1 fashion.[31] The estimated reversal potential for the EPSPs is –7 mV and –69 mV for the IPSPs (Figure 3.8). The AMPA antagonist CNQX either reduced or blocked all EPSPs. The NMDA receptor antagonist APV also reduced the amplitude of the EPSPs. When both antagonists were applied, the EPSPs were entirely eliminated (Figure 3.9). Glutamate receptor agonists and antagonists have also been shown to excite dissociated NST neurons, although it is not clear if the neurons examined in this study also include rNST neurons.[34,35] These results demonstrate that glutamate is released following ST stimulation acting at both AMPA and NMDA receptors.

3.3.2.2 *In Vivo* Studies

Although these results using brain slices demonstrate the role of glutamate as the neurotransmitter released at the primary afferent synapse in the taste pathway, they do not directly confirm a role in gustatory processing. Confirmation that glutamate mediates synaptic transmission between afferent taste fibers and NST neurons responding to stimulation of the tongue with taste stimuli resulted from *in vivo* extracellular recordings in rNST. Neurons in the rNST responding to taste stimuli isolated using extracellular recording techniques were exposed to CNQX and APV

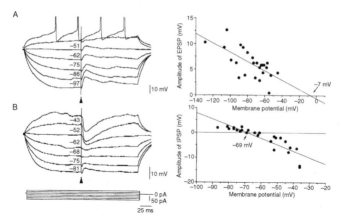

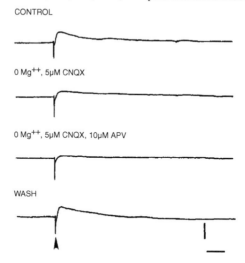

FIGURE 3.8 Postsynaptic responses recorded from two rNST neurons. (A) Voltage responses (left) to a series of hyperpolarizing current injection (bottom trace). The solitary tract was stimulated at the arrow to evoke a postsynaptic potential. The numbers above each trace indicate the membrane potential just prior to the initiation of the postsynaptic potential. This postsynaptic potential was excitatory (depolarizing), and its peak together with four other similar neurons has been plotted relative to the membrane potential and fitted with a straight line by regression analysis (right). The extrapolated reversal potential is –7 mV. (B) Similar experiment for a predominantly inhibitory postsynaptic potential. The extrapolated reversal potential is –69 mV. (Reprinted from Wang, L. and Bradley, R. M., *Brain Res.*, 702, 188, 1995, with permission from Elsevier.)

FIGURE 3.9 Average of 10 excitatory postsynaptic potentials evoked in an rNST neuron by electrical stimulation of the solitary tract before and after application of the glutamate receptor blockers, CNQX and APV. Application of both of these blockers effectively eliminates the postsynaptic potential, indicating that glutamate is the excitatory neurotransmitter between the afferent input and the rNST neurons and that both NMDA and AMPA receptors are involved. (Reprinted from Wang, L. and Bradley, R. M., *Brain Res.*, 702, 188, 1995, with permission from Elsevier.)

by microinjection. Both glutamate receptor antagonists effectively eliminated taste-evoked responses in the rNST neurons with CNQX having the predominant action, suggesting that AMPA/kainate receptors play the most significant role. Ability to block taste responses with APV was more limited, revealing a lesser role of NMDA receptors in gustatory synaptic transmission (Figure 3.10). The blocking action was the same for all tastants, indicating that all gustatory afferent fiber types employ glutamate at the first central synapse in the taste pathway. AMPA/kainate receptors also predominate at the afferent synapse in the caudal nongustatory NST.[2]

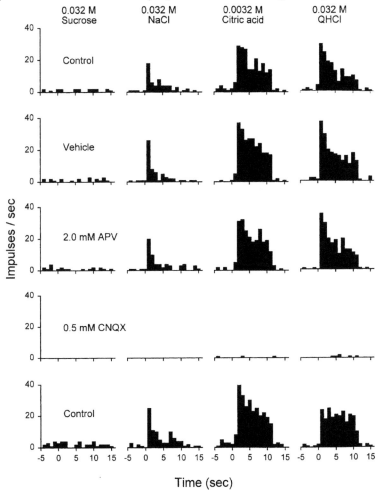

FIGURE 3.10 Responses to glutamate receptor blockers recorded *in vivo* by extracellular recording from an rNST neuron. In the control recording, taste stimuli representing the four taste qualities were flowed over the tongue and the resulting response frequency of the neuron plotted as a peristimulus-time versus action potential frequency histogram. The glutamate receptor blockers APV and CNQX were then microinjected adjacent to the neuron and the tongue stimulation sequence repeated. APV had a minimal effect on this neuron, but CNQX blocked all excitatory input from the periphery and no responses were recorded from the neuron to any taste stimuli. (From Li, C. S. and Smith, D. V., *J. Neurophysiol.*, 77, 1514, 1997, with permission.)

3.3.3 IMMUNOCYTOCHEMICAL STUDIES OF GABA IN rNST

GABA, the main inhibitory neurotransmitter in the mammalian brain, activates two distinct classes of receptors that mediate synaptic inhibition. $GABA_A$ receptors are chloride channels responsible for fast synaptic inhibition and are composed of two α, two β, and one γ subunit. The $GABA_B$ receptor is a seven-transmembrane G protein-coupled receptor that activates a second messenger system, and their response is less rapid than that following $GABA_A$ receptor activation. The GABA

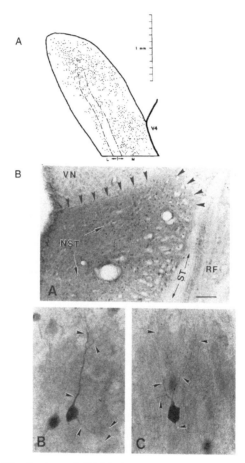

FIGURE 3.11 (A) Distribution of GABA immunoreactivity throughout the rostral and intermediate NST. Dashed lines indicate the location of the solitary tract and L and M the lateral and medial subdivisions of the NST. (Reprinted from Lasiter, P. S. and Kachele, D. L., *Brain Res. Bull.*, 21, 623, 1988, with permission from Elsevier.). (B, A) Distribution of GABA immunoreactive neurons in the rNST. The boundary of the NST is indicated by arrowheads. (B, B and B, C) Higher power photomicrographs of GABA immunoreactive neurons and dendritic processes (arrowheads). (Reprinted from Davis, B. J., *Brain Res. Bull.*, 30, 69, 1993, with permission from Elsevier.)

receptors responsible for the inhibitory activity can be distinguished by their response to pharmacological agents. The $GABA_A$ receptor is selectively sensitive to the agonist muscimol and blocked by pictrotoxin and bicuculline. $GABA_B$ receptors are activated by baclofen and antagonized by 2-OH-saclofen.

Both $GABA_A$ and $GABA_B$ receptors are found in the rNST, and recently a third GABA receptor, $GABA_C$, has been found to be transiently expressed in the rNST early in development.[36] $GABA_A$ receptors are usually localized on postsynaptic membranes, whereas $GABA_B$ receptors are found on presynaptic membranes.[37]

GABA immunoreactive neurons were shown to be present in the nongustatory NST as early as 1982[38] and subsequently demonstrated in gustatory NST, as well.[39] GABA immunoreactive neurons are distributed throughout the rNST (Figure 3.11).[39,40] Combined immunocytochemical and morphological analysis revealed that GABA is contained in the small ovoid neurons classified as local circuit or interneurons because they do not backfill when a retrograde tracer is injected into the parabrachial taste relay nucleus.[39–42]

Study of the distribution of $GABA_A$ receptor subunits has been facilitated by the availability of antibodies to the α and β subunits. The α subunit was found to be moderately abundant in the rNST,[43] whereas the β subunits are uniformly distributed throughout the rostral central subdivision with the densest labeling in the V subnucleus. $GABA_B$ receptor immunoreactivity was located primarily in neuronal somata and proximal dendrites and distributed throughout the rNST with the highest density in the medial and rostral central subdivisions (Figure 3.12).[28,44]

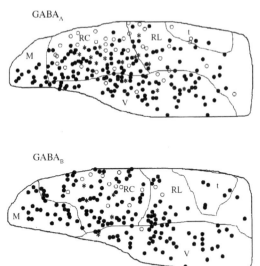

FIGURE 3.12 Diagram of the distribution of $GABA_A$, and $GABA_B$ immunoreactive neurons in the rNST. Open circles represent the location of rNST neurons that project to a parabrachial nucleus, asterisks indicate a parabrachial nucleus projecting neuron that is also immunoreactive for a GABA receptor, and filled circles are the location of immunoreactive neurons. (Reprinted from King, M. S., *Brain Res. Bull.*, 60, 241, 2003, with permission from Elsevier.)

3.3.4 Neurophysiological Studies of GABA Activity in rNST

3.3.4.1 *In Vitro* Studies

Inhibitory activity of GABA in rNST has been studied by superfusing GABA agonists and antagonists over brain slices of the medulla while recording from neurons using whole-cell patch-clamping. All rNST morphological neuron types (multipolar, ovoid, and elongate) respond to GABA.[33,45] In brain slice recordings, GABA application resulted in membrane hyperpolarization with a reduction in input resistance that was concentration dependent (Figure 3.13).[45,46] Use of the $GABA_A$ receptor agonist muscimol and $GABA_A$ receptor antagonist bicuculline confirmed that $GABA_A$ receptors were the predominant GABA receptor in rNST. A proportion of the rNST neurons was also sensitive to $GABA_B$ receptor agonists and antagonists, indicating that rNST neurons also have $GABA_B$ receptors.[45,46]

Although earlier experiments indicated that postsynaptic potentials in rNST are either depolarizing or hyperpolarizing, recent study of rNST synaptic potentials reveals that they all consist of a mixture of excitatory and inhibitory potentials. By using selective blockers of either the excitatory or inhibitory components, the different components can be studied in isolation. It has been well established using this methodology that the excitatory afferent input is mediated by glutamate, and the inhibitory transmission is mediated primarily by $GABA_A$ synapses.[47] Application of bicuculline, by eliminating the hyperpolarizing component of the PSP, revealed the excitatory component with a reversal potential

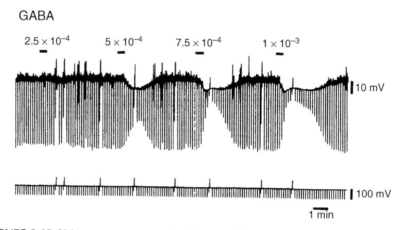

FIGURE 3.13 Voltage responses recorded from an rNST neuron in a brain slice preparation to hyperpolarizing injections of −100 pA, 100 ms current pulses. The neuron was exposed to increasing 5 sec (bar) concentrations of γ-aminobutyric acid (GABA) interspersed with control saline rinses. GABA hyperpolarizes the neuron in a concentration-dependent manner, accompanied by a decrease in input resistance. (Reprinted from Wang, L. and Bradley, R. M., *Brain Res.*, 616, 144, 1993, with permission of Elsevier.)

of –3 mV (Figure 3.14A). On the other hand, application of glutamate ionotropic receptor blockers effectively eliminates initiation of the synaptic response evoked by stimulation of the solitary tract. By increasing the stimulus strength to bypass the afferent input and activate the interneurons directly, an IPSP is elicited with a reversal potential of –88 mV (Figure 3.14B). These results confirm that all excitatory input to rNST is mediated by glutamate synapses on all neuron types in the nucleus. GABA-initiated synaptic activity results from excitation of inhibitory interneurons that synapse with other neurons and also presynaptically on the afferent terminals, as suggested by ultrastructural studies of rNST synapses.[48,49]

In addition, instead of using single shock stimuli to investigate synaptic potentials, trains of stimuli (tetanic stimulation) that mimic the frequency of afferent input to the rNST evoked by gustatory stimulation reveals additional characteristics of inhibitory activity in the rNST. Afferent gustatory input to the rNST typically consists of trains or bursts of impulses at frequencies ranging from 1 to 60 Hz,

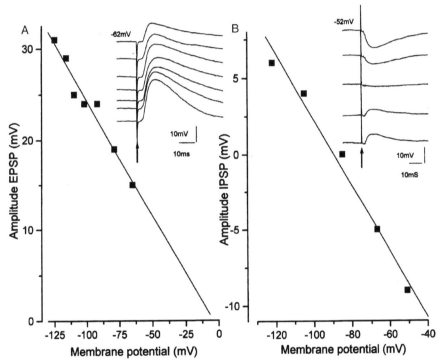

FIGURE 3.14 Voltage dependence of postsynaptic potentials elicited in an rNST neuron by electrical stimulation. (A) Voltage dependence of pure excitatory postsynaptic potentials isolated by blocking the inhibitory component with the GABA$_A$ receptor antagonist bicuculline. (B) Voltage dependence of pure inhibitory postsynaptic potentials isolated by blocking the excitatory component with the glutamate receptor antagonists CNQX and APV. In both (A) and (B), the peak amplitudes of the postsynaptic potentials are plotted against the holding potential and fitted via regression analysis. Samples of actual recordings are shown as insets. (From Grabauskas, G. and Bradley, R. M., *J. Neurophysiol.*, 76, 2919, 1996, with permission.)

depending on stimulus modality and concentration.[50-53] Tetanic stimulation at frequencies of 10–30 Hz results in sustained hyperpolarizing IPSPs that could be blocked by bicuculline. Because afferent sensory information to the rNST consists of relatively high-frequency spike trains, these short-term changes in inhibitory synaptic activity could potentially play a key role in taste processing by facilitating synaptic integration.[54]

A single tetanic stimulation results in potentiation of single stimulus shock-evoked IPSPs for a considerable time. This potentiation can last over 30 min (Figure 3.15).

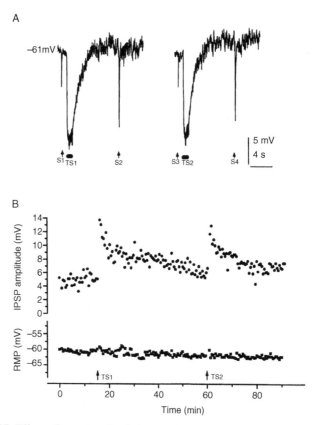

FIGURE 3.15 Effect of tetanic stimulation on inhibitory postsynaptic potentials (IPSPs) recorded from an rNST neuron in a brain slice preparation. IPSPs were isolated by chronic application of glutamate receptor blockers and elicited by electrical stimulation of the rNST. (A) After eliciting a control IPSP (S1), a tetanic stimulation (TS1) evoked potentiation of the amplitude of a subsequent IPSP (S2). 45 min after the initial potentiation, the IPSP returns to control levels (S3). A second tetanic stimulation (TS2) results in potentiation again (S4). (B) Tetanic stimulation (TS1) potentiates the IPSP amplitude (•) for a considerable length of time. Tetanic stimulation does not change the resting potential (■). Once the IPSP returns to control levels, a second tetanic stimulation (TS2) results in potentiation again. (Reprinted from Grabauskas, G. and Bradley, R. M., *Neuroscience*, 94, 1173, 1999, with permission of Elsevier.)

Tetanic stimulation results in activation of all postsynaptic $GABA_A$ receptors and induces long-lasting changes in the presynaptic GABAergic neuron. These long-lasting changes of the presynaptic neuron facilitate the release of GABA during single-stimulus shock stimuli, and as a consequence, more postsynaptic receptors are activated during a single-stimulus shock. Long-term potentiation of inhibition in the gustatory relay nucleus has the capability to profoundly alter transmission of taste information and, therefore, may be of significance in the mechanism of taste-related learning phenomena.[55] A further property of the rNST IPSPs is that the synaptic strength of the inhibitory synapses is modified by prior activity. Specifically, the amplitude and decay time of the IPSPs depend on previous synaptic activity, resulting in accumulation of transmitter in the synaptic cleft.[56,57]

Neural activity resulting from gustatory stimulation of the tongue travels to the rNST in both the chorda tympani from the anterior tongue and the glossopharyngeal nerve from the posterior tongue. A number of investigators have demonstrated that convergent interactions occur among taste buds on the anterior and posterior tongue at the level of the rNST.[58–61] The neurobiology of these synaptic interactions was investigated in horizontal brain slices by electrically stimulating the central projection area of both the chorda tympani and glossopharyngeal nerves and recording PSPs from single neurons. The resulting synaptic potential was a sum of the two individual potentials. If stimulation of the rostral and intermediate sites both elicited depolarizing PSPs, then when both sites were stimulated simultaneously, the resulting PSP was an arithmetic sum of the two individual PSPs.

In contrast, IPSPs evoked by stimulation of the rostral and intermediate sites also summed, but the summation was not linear. When the PSP evoked by stimulation at one site was excitatory but inhibitory at the other site, the PSP waveform resulting from simultaneous stimulation was a complex mixture of the two potentials. Moreover, the inhibitory potential was capable of blocking action potentials evoked by stimulating the excitatory site (Figure 3.16).[47] These results dramatically illustrate the complexity of processing possible by the rNST integrating gustatory input from tongue chemoreceptors. Moreover, input from somatosensory receptors in the oral cavity will also contribute to this complexity, as stimulation is a dynamic interaction of sensory modalities.[62]

The postnatal development of synaptic inhibition in rNST has also been examined.[36] All IPSPs recorded in postnatal rats were hyperpolarizing, but the rise and decay time constants of the single-stimulus shock-evoked IPSPs decreased, and the inhibition response-concentration function to the $GABA_A$ receptor antagonist bicuculline shifted to the left during development (Figure 3.17A). In postnatal animals aged 0–14 days, but not older animals, the IPSPs had a bicuculline-insensitive component that was sensitive to block by picrotoxin, suggesting a transient expression of $GABA_C$ receptors (Figure 3.17B). These developmental changes in inhibitory synaptic activity may operate to shape synaptic activity in early development of the rat gustatory system during a time of maturation of taste preferences and aversions.[36]

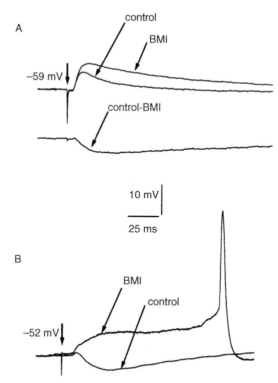

FIGURE 3.16 Demonstration that postsynaptic potentials in the rNST have both excitatory and inhibitory components. (A) Average of 10 depolarizing postsynaptic potentials before (control) and after application of the $GABA_A$ receptor blocker bicuculline. The postsynaptic potential increases in amplitude once the inhibitory component is blocked. The inhibitory component is revealed by subtracting the control record from the record after bicuculline application (control-BMI). (B) Similar experiment on a predominantly inhibitory postsynaptic potential (control). Once the inhibitory component is blocked with bicuculline, the excitatory component was revealed and stimulation now elicits an action potential. (From Grabauskas, G. and Bradley, R. M., *J. Neurophysiol.*, 76, 2919, 1996, with permission.)

3.3.4.2   *In Vivo* Studies

The brainstem slice experiments suggest that rNST neurons are under tonic GABAergic inhibition. This was directly tested *in vivo* on taste-responsive neurons in which the spontaneous activity was suppressed in a concentration-dependent manner by microinjection of GABA.[63] Responses of the rNST neurons to tongue-applied taste stimuli were also enhanced by microinjection of the $GABA_A$ receptor antagonist bicuculline. The fact that rNST neurons are chronically inhibited implies that by decreasing inhibition, the neurons would be more excitable or that by increasing inhibition they would be less excitable. Thus, the level of tonic GABA release has the ability to modulate the excitability of rNST neurons and, therefore, alter their response properties.

Injection of the benzodiazepine (BZ) receptor agonist diazepam (Valium) into the 4th ventricle enhances food palatability and intake,[64] and it has been hypothesized

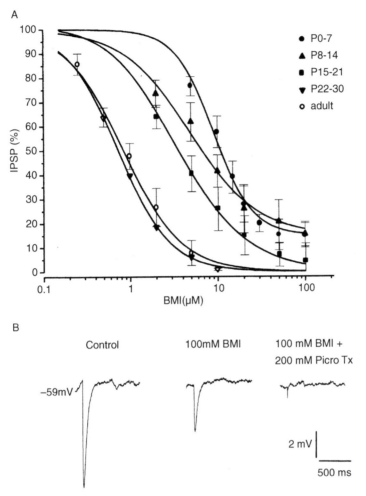

FIGURE 3.17 (A) Effect of different concentrations of the GABA$_A$ antagonist bicuculline (BMI) on the amplitude of single shock-evoked inhibitory postsynaptic potentials at different rat postnatal ages. The inhibition-response curves shift to the left with age, indicating a development change in sensitivity to bicuculline. (B) Single-stimulus shock-evoked inhibitory postsynaptic potentials are not completely blocked by bicuculline application. The bicuculline-insensitive component was blocked by application of picrotoxin (PicroTx), a blocker of GABA$_C$ receptors. (From Grabauskas, G. and Bradley, R. M., *J. Neurophysiol.*, 85, 2203, 2001, with permission.)

that this phenomenon results from the action of diazepam on rNST neurons. Benzodiazepines facilitate GABA-induced IPSPs in postsynaptic neurons that have GABA$_A$ receptors by modulating Cl$^-$ channel activity. In brain slices maintained at body temperature, pharmacologically isolated IPSPs are diazepam sensitive, demonstrating that rNST neurons express GABA$_A$ receptors with BZ-binding sites. It is possible, therefore, that application of BZ into the 4th ventricle influences food intake by potentiation of GABAergic neurotransmission in the rNST.[65]

3.4 OTHER NEUROTRANSMITTERS INVOLVED IN SYNAPTIC TRANSMISSION IN rNST

3.4.1 SUBSTANCE P

3.4.1.1 Immunocytochemical Investigation

Several other neurotransmitters and receptors have been localized within the rNST.[66] The most-studied neuropeptide in the rNST is the tachykinin, substance P (SP). The majority of SP labeling within the rNST is in fibers and puncta, with very few neurons expressing this peptide,[67-69] suggesting that the source of SP is extranuclear. As the solitary tract includes some SP-IR fibers[67] and the petrosal ganglion contains neurons immunoreactive for this peptide,[12] some SP may be supplied by afferent nerve fibers. However, the density of SP-labeled fibers and puncta within the rNST suggests other sources, as well. Therefore, it is probable that forebrain structures that project to the rNST like the insular cortex, bed nucleus of the stria terminalis, central nucleus of the amygdala, and lateral hypothalamus that contain SP-IR somata provide some of this input, as well.[69] In rat, labeling for SP and its preferred receptor (NK-1) decreases in density when moving medial to lateral within the rNST.[67,70] There is moderate to heavy labeling for SP and NK-1 in RC, M, and V subnuclei, implying that this peptide could regulate responses to gustatory input as well as the ascending and descending outputs of the rNST.

3.4.1.2 Neurophysiological Investigation

Application of SP to neurons of the rNST in an *in vitro* brain slice preparation depolarized the majority of the neurons studied (68%) in a dose-dependent manner.[71] SP excited all three of the morphological groups (multipolar, elongate, and ovoid), indicating that SP potentially plays an important role in gustatory processing by the rNST. The excitatory effect of SP was the result of direct postsynaptic action on the neurons.[71] Similar results were reported when SP was microinjected onto rNST neurons in an *in vivo* preparation.[72] In this study, substance P excited 48% of the neurons and enhanced the response of the rNST neurons to stimulation of the tongue with taste stimuli. In both the *in vitro* and *in vivo* studies, a small number of neurons were inhibited by substance P application, demonstrated by membrane hyperpolarization or suppression of taste responses, respectively. Thus, substance P effectively modulates the activity of gustatory neurons in rNST.

Another tachykinin (neurokinin B) and its preferred receptor (NK-3) also are present within the rNST.[67,73,74] Due to the intense labeling of NK-3 receptors within the glossopharyngeal terminal field,[67] it is possible that these receptors modulate the transmission of information from the posterior tongue to the rNST.

3.4.2 OTHER NEUROTRANSMITTERS AND NEUROMODULATORS

3.4.2.1 Opioids

In the rNST, there is an extensive plexus of nerve fibers and terminal swellings expressing methionine enkephalin (met-enkephalin) immunoreactivity as well as some enkephalin-reactive somata.[68] Microinjections of methionine enkephalin have been shown to modulate the activity of gustatory neurons in rNST.[75] During extracellular recording from rNST neurons, met-enkephalin was microinjected, resulting in the suppression of taste-initiated responses in 24% of the recorded neurons. The opioid antagonist naltrexone effectively blocked the inhibitory effect of met-enkephalin, suggesting that opiates maintain a tonic inhibitory action on a subpopulation of taste-responsive neurons in rNST.

3.4.2.2 Cholecystokinin

Cholecystokinin (CCK), a peptide secreted by gut lining cells, plays an important role in digestion. CCK has also been shown to cause satiety by acting centrally,[76] and neurons immunoreactive for CCK have been located in the NST[29,77] and the PBN, including the waist area.[77–81] Thus, CCK is present in the NST and taste-responsive area of the PBN, suggesting that this neuropeptide may have a role in taste processing. Although investigators using electrophysiological methods have demonstrated that CCK can modulate the excitatory response initiated by vagal stimulation,[82] there are no similar studies on taste-responsive neurons in the rNST.

3.4.2.3 Dopamine

Dopaminergic neurons in the rNST, identified as immunoreactive for tyrosine hydroxylase but not dopamine β-hydroxylase, tend to be relatively large and located within the RC subnucleus,[9,83] suggesting that they should receive input from cranial nerves and contribute to the ascending pathways (Figure 3.18). In fact, many dopaminergic

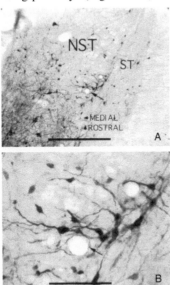

FIGURE 3.18 Tyrosine hydroxylase immunoreactive neurons in rNST. (A) Horizontal section through the rNST showing that tyrosine hydroxylase immunoreactive neurons are located medial to the solitary tract (ST). (B) Higher magnification image of tyrosine hydroxylase immunoreactive neurons and dendrites. The arrow in B indicates the location of the neurons A. (From Davis, B. J., *J. Comp. Neurol.*, 392, 78, 1998, with permission.)

neurons have been shown to receive input from the chorda tympani nerve[83] and to respond to the oral application of tastants.[84] Implying a function for dopamine within the rNST, an autoradiographic study indicates that D2, but not D1, receptors are located within the nucleus.[29]

3.4.2.4 Acetylcholine

Although abundant in the caudal NST,[85] neurons that produce acetylcholine have not been identified in the rNST. Moreover, the cholinergic neurons that have been identified in the caudal NST[86] resemble parasympathetic preganglionic ectopic neurons labeled following applications of retrograde tracers[87] and may not, therefore, be true NST neurons. Cholinergic receptors have been demonstrated in the rNST that were speculated to be involved with gustatory function[88] and have recently been identified on rNST neurons in brain slice experiments.[89] The discovery of intense staining for acetylcholinesterase in gustatory nerve terminal fields also prompted the suggestion of a possible role for cholinergic transmission in taste processing.[90] However, in a later study, this idea was tested and rejected, leaving the role of acetylcholinesterase immunostaining unclear.[91] Thus, at present, cholinergic neurotransmission in rNST gustatory processing remains to be defined.

3.4.2.5 Nitric Oxide, Neuropeptide Y, Somatostatin, Serotonin, and Histamine

Based on immunochemical studies, a number of other putative neurotransmitters and neuromodulators have been identified in rNST. Nitric oxide immunoreactivity is present throughout the rostral-caudal extent of the NST,[92] often colocalized with glutamate-containing neurons.[93,94] Many neurons within the RL subnucleus surrounding the solitary tract stain intensely for NADPH-diaphorase, suggesting that nitric oxide levels within the rNST may be regulated by orosensory input.[95] Other peptides located within the rNST include neuropeptide Y and somatostatin,[66,92] but little is known about their distribution or function within the nucleus. Serotonin[96] and histamine[97] receptors also have been localized within the rNST, but definitive sources of these neurotransmitters and their functions remain unknown. Further neuropharmacological investigations of the roles of all these neuropeptides in gustatory processing remains to be accomplished.

3.5 FUTURE DIRECTIONS

The essence of neural processing involves synaptic interactions, and synaptic interactions involve neurotransmitters. Investigators studying taste processing have only recently turned their attention to the role of neurotransmitters at each relay in the taste pathway. In the review by Travers et al. in 1987[98] on gustatory neural processing in the hindbrain, neurotransmitters are not even mentioned. Since that time, a number of investigators have demonstrated both anatomically and physiologically the role neurotransmission plays in taste processing. They have determined that glutamate is the excitatory transmitter between afferent taste fibers and neurons of the rNST.

It has been shown that inhibition mediated principally by $GABA_A$ receptors is an important factor in modulating the afferent input, and that neurons of the rNST are under tonic GABA inhibition. Moreover, a form of long-term inhibition must also be factored into gustatory processing by rNST circuits. Although glutamate and GABA play a pivotal role in synaptic transmission in the rNST, afferent information is distributed to a number of destinations after arriving at the brainstem, and these circuits may use different neurotransmitters. Although glutamate and GABA have received much attention, other neurotransmitters are also important as well, and some of these have been characterized, particularly substance P. Besides neurotransmitters active in the ascending taste pathway, other neurotransmitters are found in pathways descending to the rNST from more rostral brain areas. Although much more information remains to be acquired, it is evident from the current state of knowledge that the apparent complexity of synaptic interactions at all levels of the taste pathway should be taken into account when the subject of central taste coding is discussed.

ACKNOWLEDGMENT

The preparation of this chapter was supported in part by NIH grant DC 000288 from the National Institute on Deafness and Other Communication Disorders to R. M. Bradley.

REFERENCES

1. Van Giersbergen, P. L. M., Palkovits, M. and de Jong, W. Involvement of neurotransmitters in the nucleus tractus solitarii in cardiovascular regulation, *Physiol. Rev.*, 72, 789, 1992.
2. Andresen, M. C. and Kunze, D. L. Nucleus tractus solitarius—Gateway to neural circulatory control, *Annu. Rev. Physiol.*, 56, 93, 1994.
3. Lawrence, A. J. and Jarrott, B. Neurochemical modulation of cardiovascular control in the nucleus tractus solitarius, *Prog. Neurobiol.*, 48, 21, 1996.
4. Bradley, R. M., King, M. S., Wang, L. and Shu, X. Q. Neurotransmitter and neuromodulatory activity in the gustatory zone of the nucleus tractus solitarius, *Chem. Senses*, 21, 377, 1996.
5. Bradley, R. M. and Grabauskas, G. Neural circuits for taste: excitation, inhibition and synaptic plasticity in the rostral gustatory zone of the nucleus of the solitary tract, in *Olfaction and Taste XII*, Murphy, C. (Ed.), New York Academy of Sciences, New York, 1998, p. 467.
6. Smith, D. V., Li, C.-S. and Davis, B. J. Excitatory and inhibitory modulation of taste responses in the hamster brainstem, in *Olfaction and Taste XII*, Murphy, C. (Ed.), New York Academy of Sciences, New York, 1998, p. 450.
7. Ambalavanar, R., Ludlow, C. L., Wenthold, R. J., Tanaka, Y., Damirjian, M. and Petralia, R. S. Glutamate receptor subunits in the nucleus of the tractus solitarius and other regions of the medulla oblongata in the cat, *J. Comp. Neurol.*, 402, 75, 1998.
8. Saha, S., Sieghart, W., Fritschy, J. M., McWilliam, P. N. and Batten, T. F. gamma-aminobutyric acid receptor ($GABA_A$) subunits in rat nucleus tractus solitarii (NTS) revealed by polymerase chain reaction (PCR) and immunohistochemistry, *Mol. Cell. Neurosci.*, 17, 241, 2001.

9. Davis, B. J. and Jang, T. Tyrosine hydroxylase-like and dopamine beta-hydroxylase-like immunoreactivity in the gustatory zone of the nucleus of the solitary tract in the hamster: light- and electron-microscopic studies, *Neuroscience*, 27, 949, 1988.

10. Czyzyk-Krzeska, M. F., Bayliss, D. A., Seroogy, K. B. and Millhorn, D. E. Gene expression for peptides in neurons of the petrosal and nodose ganglia in rat, *Exp. Brain Res.*, 83, 411, 1991.

11. Finley, J. C. W., Polak, J. and Katz, D. M. Transmitter diversity in carotid body afferent neurons: dopaminergic and peptidergic phenotypes, *Neuroscience*, 51, 973, 1992.

12. Helke, C. J. and Hill, K. M. Immunohistochemical study of neuropeptides in vagal and glossopharyngeal afferent neurons in the rat, *Neuroscience*, 26, 539, 1988.

13. Helke, C. J. and Niederer, A. J. Studies on the coexistence of substance P with other putative neurotransmitters in the nodose and petrosal ganglia, *Synapse*, 5, 144, 1990.

14. Helke, C. J. and Rabchevsky, A. Axotomy alters putative neurotransmitters in visceral sensory neurons of the nodose and petrosal ganglia, *Brain Res.*, 551, 44, 1991.

15. Ichikawa, H., Jacobowitz, D. M., Winsky, L. and Helke, C. J. Calretinin-immunoreactivity in vagal and glossopharyngeal sensory neurons of the rat: distribution and coexistence with putative transmitter agents, *Brain Res.*, 557, 316, 1991.

16. Okada, J. and Miura, M. Transmitter substances contained in the petrosal ganglion cells determined by a double-labeling method in the rat, *Neurosci. Lett.*, 146, 33, 1992.

17. Caicedo, A., Zucchi, B., Pereira, E. and Roper, S. D. Rat gustatory neurons in the geniculate ganglion express glutamate receptor subunits, *Chem. Senses*, 29, 463, 2004.

18. King, M. S. and Bradley, R. M. Biophysical properties and responses to glutamate receptor agonists of identified subpopulations of rat geniculate ganglion neurons, *Brain Res.*, 866, 237, 2000.

19. Koga, T. and Bradley, R. M. Biophysical properties and responses to neurotransmitters of petrosal and geniculate ganglion neurons innervating the tongue, *J. Neurophysiol.*, 84, 1404, 2000.

20. Mayer, M. L. and Westbrook, G. L. The physiology of excitatory amino acids in the vertebrate central nervous system, *Prog. Neurobiol.*, 28, 197, 1987.

21. Hollmann, M. and Heinemann, S. Cloned glutamate receptors, *Annu. Rev. Neurosci.*, 17, 31, 1994.

22. Dingledine, R., Borges, K., Bowie, D. and Traynelis, S. F. The glutamate receptor ion channels, *Pharmacol. Rev.*, 51, 7, 1999.

23. Cincotta, M., Beart, P. M., Summers, R. J. and Lodge, D. Bidirectional transport of NMDA receptor and ionophore in the vagus nerve, *Eur. J. Phamacol.*, 160, 167, 1989.

24. Aicher, S. A., Sharma, S. and Pickel, V. M. *N*-methyl-D-aspartate receptors are present in vagal afferents and their dendritic targets in the nucleus tractus solitarius, *Neuroscience*, 91, 119, 1999.

25. MacDermott, A. B., Role, L. W. and Siegelbaum, S. A. Presynaptic ionotropic receptors and the control of transmitter release, *Annu. Rev. Neurosci.*, 22, 443, 1999.

26. Gill, C. F., Madden, J. M., Roberts, B. P., Evans, L. D. and King, M. S. A subpopulation of neurons in the rat rostral nucleus of the solitary tract that project to the parabrachial nucleus express glutamate-like immunoreactivity, *Brain Res.*, 821, 251, 1999.

27. Sweazey, R. D. Distribution of aspartate and glutamate in the nucleus of the solitary tract of the lamb, *Exp. Brain Res.*, 105, 241, 1995.

28. King, M. S. Distribution of immunoreactive GABA and glutamate receptors in the gustatory portion of the nucleus of the solitary tract in rat, *Brain Res. Bull.*, 60, 241, 2003.

29. Qian, M., Johnson, A. E., Källström, L., Carrer, H. and Södersten, P. Cholecystokinin, dopamine D_2 and *N*-methyl-D-aspartate binding sites in the nucleus of the solitary tract of the rat: possible relationship to ingestive behavior, *Neuroscience*, 77, 1077, 1997.

30. Li, C. S. and Smith, D. V. Glutamate receptor antagonists block gustatory afferent input to the nucleus of the solitary tract, *J. Neurophysiol.*, 77, 1514, 1997.

31. Wang, L. and Bradley, R. M. *In vitro* study of afferent synaptic transmission in the rostral gustatory zone of the rat nucleus of the solitary tract, *Brain Res.*, 702, 188, 1995.

32. Du, J. and Bradley, R. M. Electrophysiological and morphological properties of neurons acutely isolated from the rostral gustatory zone of the rat nucleus of the solitary tract, *Chem. Senses*, 21, 729, 1996.

33. Du, J. and Bradley, R. M. Effect of GABA on acutely isolated neurons from the gustatory zone of the rat nucleus of the solitary tract, *Chem. Senses*, 23, 683, 1998.

34. Nakagawa, T., Shirasaki, T., Wakamori, M., Fukuda, A. and Akaike, N. Excitatory amino acid response in isolated nucleus tractus solitarii neurons of the rat, *Neurosci. Res.*, 8, 114, 1990.

35. Nakagawa, T., Shirasaki, T., Tateishi, N., Murase, K. and Akaike, N. Effects of antagonists on *N*-methyl-D-aspartate response in acutely isolated nucleus tractus solitarii neurons of the rat, *Neurosci. Lett.*, 113, 169, 1990.

36. Grabauskas, G. and Bradley, R. M. Postnatal development of inhibitory synaptic transmission in the rostral nucleus of the solitary tract, *J. Neurophysiol.*, 85, 2203, 2001.

37. Cherubini, E. and Strata, F. GABA$_c$ receptors: a novel receptor family with unusual pharmacology, *News Physiol. Sci.*, 12, 136, 1997.

38. Dietrich, W. D., Lowry, O. H. and Loewy, A. D. The distribution of glutamate, GABA and aspartate in the nucleus tractus solitarius of the cat, *Brain Res.*, 237, 254, 1982.

39. Lasiter, P. S. and Kachele, D. L. Organization of GABA and GABA-transaminase containing neurons in the gustatory zone of the nucleus of the solitary tract, *Brain Res. Bull.*, 21, 623, 1988.

40. Davis, B. J. GABA-like immunoreactivity in the gustatory zone of the nucleus of the solitary tract in the hamster: light- and electron-microscopic studies, *Brain Res. Bull.*, 30, 69, 1993.

41. Leonard, N. L., Renahan, W. E. and Schweitzer, L. Structure and function of gustatory neurons in the nucleus of the solitary tract. IV. The morphology and synaptology of GABA-immunoreactive terminals, *Neuroscience*, 92, 151, 1999.

42. Sweazey, R. D. Distribution of GABA and glycine in the lamb nucleus of the solitary tract, *Brain Res.*, 737, 275, 1996.

43. Hironaka, T., Morita, Y., Hagihira, S., Tateno, E., Kita, H. and Tohyama, M. Localization of GABA$_A$-receptor α_1 subunit mRNA-containing neurons in the lower brainstem of the rat, *Mol. Brain Res.*, 7, 335, 1990.

44. Terai, K., Tooyama, I. and Kimura, H. Immunohistochemical localization of GABA$_A$ receptors in comparison with GABA-immunoreactive structures in the nucleus tractus solitarii of the rat, *Neuroscience*, 82, 843, 1998.

45. Wang, L. and Bradley, R. M. Influence of GABA on neurons of the gustatory zone of the rat nucleus of the solitary tract, *Brain Res.*, 616, 144, 1993.

46. Liu, H., Behbehani, M. M. and Smith, D. V. The influence of GABA on cells in the gustatory region of the hamster solitary nucleus, *Chem. Senses*, 18, 285, 1993.

47. Grabauskas, G. and Bradley, R. M. Synaptic interactions due to convergent input from gustatory afferent fibers in the rostral nucleus of the solitary tract, *J. Neurophysiol.*, 76, 2919, 1996.

48. Whitehead, M. C. Anatomy of the gustatory system in the hamster: synaptology of facial afferent terminals in the solitary nucleus, *J. Comp. Neurol.*, 244, 72, 1986.

49. Brining, S. K. and Smith, D. V. Distribution and synaptology of glossopharyngeal afferent nerve terminals in the nucleus of the solitary tract of the hamster, *J. Comp. Neurol.*, 365, 556, 1996.

50. Fishman, I. Y. Single fiber gustatory impulses in rat and hamster, *J. Cell Comp. Physiol.*, 49, 319, 1957.

51. Frank, M. E., Bieber, S. L. and Smith, D. V. The organization of taste sensibilities in hamster chorda tympani nerve fibers, *J. Gen. Physiol.*, 91, 861, 1988.

52. Ogawa, H., Sato, M. and Yamashita, S. Multiple sensitivity of chorda tympani fibers of the rat and hamster to gustatory and thermal stimuli, *J. Physiol.*, 199, 223, 1968.

53. Ogawa, H., Sato, M. and Yamashita, S. Variability in impulse discharges in rat chorda tympani fibers in response to repeated gustatory stimulations, *Physiol. Behav.*, 11, 469, 1973.

54. Grabauskas, G. and Bradley, R. M. Tetanic stimulation induces short-term potentiation of inhibitory synaptic activity in the rostral nucleus of the solitary tract, *J. Neurophysiol.*, 79, 595, 1998.

55. Grabauskas, G. and Bradley, R. M. Potentiation of GABAergic synaptic transmission in the rostral nucleus of the solitary tract, *Neuroscience*, 94, 1173, 1999.

56. Grabauskas, G. and Bradley, R. M. Frequency-dependent properties of inhibitory synapses in the rostral nucleus of the solitary tract, *J. Neurophysiol.*, 89, 199, 2003.

57. Grabauskas, G. Time course of GABA in the synaptic clefts of inhibitory synapses in the rostral nucleus of the solitary tract, *Neurosci. Lett.*, 373, 10, 2005.

58. Halpern, B. P. and Nelson, L. M. Bulbar gustatory responses to anterior and to posterior tongue stimulation in the rat, *Am. J. Physiol.*, 209, 105, 1965.

59. Kveton, J. F. and Bartoshuk, L. M. The effect of unilateral chorda tympani damage on taste, *Laryngoscope*, 104, 25, 1994.

60. Sweazey, R. D. and Smith, D. V. Convergence onto hamster medullary taste neurons, *Brain Res.*, 408, 173, 1987.

61. Lehman, C. D., Bartoshuk, L. M., Catalanotto, F. C., Kveton, J. F. and Lowlicht, R. A. Effect of anesthesia of the chorda tympani nerve on taste perception in humans, *Physiol. Behav.*, 57, 943, 1995.

62. Green, B. G. Studying taste as a cutaneous sense, *Food Qual. Pref.*, 14, 99, 2002.

63. Smith, D. V. and Li, C.-S. Tonic GABAergic inhibition of taste-responsive neurons in the nucleus of the solitary tract, *Chem. Senses*, 23, 159, 1998.

64. Berridge, K. C. and Peciña, S. Benzodiazepines, appetite, and taste palatability, *Neurosci. Biobehav. Rev.*, 19, 121, 1995.

65. Grabauskas, G. and Bradley, R. M. Effect of diazepam on inhibitory postsynaptic potentials in nucleus of the solitary tract of neonatal rats is temperature dependent. *Chemical Senses* 25, 678, 2000.

66. Mantyh, P. W. and Hunt, S. P. Neuropeptides are present in projection neurons at all levels in visceral taste pathways: from periphery to sensory cortex, *Brain Res.*, 299, 297, 1984.

67. Harrison, T. A., Hoover, D. B. and King, M. S. Distinct regional distributions of NK1 and NK3 neurokinin receptor immunoreactivity in rat brainstem gustatory centers, *Brain Res. Bull.*, 63, 7, 2004.

68. Davis, B. J. and Kream, R. M. Distribution of tachykinin- and opioid-expressing neurons in the hamster solitary nucleus: an immuno- and *in situ* hybridization histochemical study, *Brain Res.*, 616, 6, 1993.

69. Ljungdahl, A., Hokfelt, T. and Nilsson, G. Distribution of substance P-like immunoreactivity in the central nervous system of the rat. I. Cell bodies and nerve terminals, *Neuroscience*, 3, 861, 1978.

70. Davis, B. J. and Smith, H. M. Neurokinin-1 receptor immunoreactivity in the nucleus of the solitary tract in the hamster, *NeuroReport*, 10, 1003, 1999.

71. King, M. S., Wang, L. and Bradley, R. M. Substance P excites neurons in the gustatory zone of the rat nucleus tractus solitarius, *Brain Res.*, 619, 120, 1993.

72. Davis, B. J. and Smith, D. V. Substance P modulates taste responses in the nucleus of the solitary tract of the hamster, *NeuroReport*, 8, 1723, 1997.

73. Lucas, L. R., Hurley, D. L., Krause, J. E. and Harlan, R. E. Localization of the tachykinin neurokinin B precusor peptide in rat brain by immunocytochemistry and *in situ* hybridization, *Neuroscience*, 51, 317, 1992.

74. Merchenthaler, I., Maderdrut, J. L., O'Harte, F. and Conlon, J. M. Localization of neurokinin B in the central nervous system of the rat, *Peptides*, 13, 815, 1992.

75. Li, C. S., Davis, B. J. and Smith, D. V. Opioid modulation of taste responses in the nucleus of the solitary tract, *Brain Res.*, 965, 21, 2003.

76. Baile, C. A., McLaughlin, C. L. and Della-Fera, M. A. Role of cholecystokinin and opioid peptides in control of food intake, *Physiol. Rev.*, 66, 172, 1986.

77. Kubota, Y., Inagaki, S., Shiosaka, S., Cho, H. J., Tateishi, K., Hasimura, E., Hamaoka, T. and Tohyama, M. The distribution of cholecystokinin octapeptide-like structures in the lower brain stem of the rat: an immunohistochemical analysis, *Neuroscience*, 9, 587, 1983.

78. Block, C. H. and Hoffman, G. Neuropeptides and monoamine components of the parabrachial pontine complex, *Peptides*, 8, 267, 1987.

79. Sutin, E. L. and Jacobowitz, D. M. Immunocytochemical localization of peptides and other neurochemicals in the rat laterodorsal nucleus and adjacent area, *J. Comp. Neurol.*, 270, 243, 1988.

80. Herbert, H. and Saper, C. B. Cholecystokinin-, galanin-, and corticotropin-releasing factor-like immunoreactive projections from the nucleus of the solitary tract to the parabrachial nucleus in the rat, *J. Comp. Neurol.*, 293, 581, 1990.

81. Födor, M., Gorcs, T. J. and Plakovits, M. Immunohistochemical study on the distribution of neuropeptides within the pontine tegmentum—particularly the parabrachial nucleus and the locus coeruleus of the human brain, *Neuroscience*, 46, 891, 1992.

82. Saleh, T. M. and Cechetto, D. F. Peptide changes in the parabrachial nucleus following cervical vagal stimulation, *J. Comp. Neurol.*, 366, 390, 1996.

83. Davis, B. J. Synaptic relationships between the chorda tympani and tyrosine hydroxylase-immunoreactive dendritic processes in the gustatory zone of the nucleus of the solitary tract in the hamster, *J. Comp. Neurol.*, 392, 78, 1998.

84. Davis, B. J. and Smith, H. M. Taste-induced Fos expression in dopaminergic neurons in the nucleus of the solitary tract in hamster. *Chemical Senses* 24, 597, 1999.

85. Maley, B. E. and Seybold, V. S. Distribution of [^{3}H]quinuclidinyl benzilate, [^{3}H]nicotine, and [^{125}I]alpha-bungarotoxin binding sites in the nucleus tractus solitarii of the cat, *J. Comp. Neurol.*, 327, 194, 1993.

86. Armstrong, D. M., Rotler, A., Hersh, L. B. and Pickel, V. M. Localization of choline acetyltransferase in perikarya and dendrites within the nuclei of the solitary tracts, *J. Neurosci. Res.*, 20, 279, 1988.

87. Fox, E. A. and Powley, T. L. Morphology of identified preganglionic neurons in the dorsal motor nucleus of the vagus, *J. Comp. Neurol.*, 322, 79, 1992.

88. Wamsley, J. K., Lewis, M. S., Young, W. S. and Kuhar, M. J. Autoradiographic localization of muscarinic cholinergic receptors in rat brainstem, *J. Neurosci.*, 1, 176, 1981.

89. Uteshev, V. V., Bryant, J. and Smith, D. V. Functional and morphological heterogeneity of neurons in the rostral nucleus of the solitary tract. *Abstract Viewer/Itinerary Planner* 280, 17, 2005.

90. Barry, M. A., Halsell, C. B. and Whitehead, M. C. Organization of the nucleus of the solitary tract in the hamster: acetylcholinesterase, NADH dehydrogenase, and cytochrome oxidase histochemistry, *Microsc. Res. Tech.*, 26, 231, 1993.

91. Barry, M. A., Haglund, S. and Savoy, L. D. Association of extracellular acetylcholinesterase with gustatory nerve terminal fibers in the nucleus of the solitary tract, *Brain Res.*, 921, 12, 2002.

92. Lawrence, A. J., Castillo-Meléndez, M., McLean, K. J. and Jarrott, B. The distribution of nitric oxide synthase-, adenosine deaminase- and neuropeptide Y-immunoreactivity through the entire rat nucleus tractus solitarius — effect of unilateral nodose ganglionectomy, *J. Chem. Neuroanat.*, 15, 27, 1998.

93. Lin, L. H. and Talman, W. T. Colocalization of GluR1 and neuronal nitric oxide synthase in rat nucleus tractus solitarii neurons, *Neuroscience*, 106, 801, 2002.

94. Lin, L. H. and Talman, W. T. Nitroxidergic neurons in rat nucleus tractus solitarii express vesicular glutamate transporter 3, *J. Chem. Neuroanat.*, 29, 179, 2005.

95. Takemura, M., Wakisaka, S., Yoshida, A., Nagase, Y., Bae, Y. C. and Shigenaga, Y. NADPH-diaphorase in the spinal trigeminal nucleus oralis and rostral solitary tract nucleus of rats, *Neuroscience*, 61, 587, 1994.

96. Thor, K. B., Blitz-Siebert, A. and Helke, C. J. Autoradiographic localization of $5HT_1$ binding sites in the medulla oblongata of the rat, *Synapse*, 10, 185, 1992.

97. Cumming, P., Gjedde, A. and Vincent, S. Histamine H_3 binding sites in rat brain: localization in the nucleus of the solitary tract, *Brain Res.*, 641, 198, 1994.

98. Travers, J. B., Travers, S. P. and Norgren, R. Gustatory neural processing in the hindbrain, *Annu. Rev. Neurosci.*, 10, 595, 1987.

4 Reflex Connections

Robert M. Bradley and Miwon Kim

CONTENTS

4.1 INTRODUCTION

The NST plays a pivotal role as a portal of entry of visceral and sensory information arising from the gut, cardiorespiratory, somatosensory, and taste systems. The role of the NST in autonomic and circulatory control has been recently reviewed[1,2] but connections from the NST to brainstem motor systems responsible for muscle activity related to feeding and salivary secretion have only recently received attention. Studies of the neurons and synaptic connections between taste afferent input and preganglionic neurons controlling the salivary glands are limited. And although there is considerable anatomical knowledge of the circuits connecting the rostral NST (rNST) to the orofacial motor neuron pools, details of the neurobiology of the synaptic connections remain to be studied. Despite this paucity of information, some functional properties of the reflex connections between the rNST and brainstem motor and preganglionic secretomotor neurons are known and are reviewed here.

Gustatory information processed by the rNST is distributed to a number of brain locations. By far the most studied is the rostral connection via the parabrachial nucleus to higher brain centers responsible for sensory and hedonic gustatory processing. However, just as important are connections to brainstem loci that are the basis of a number of reflex activities.[3,4] During feeding, several oral-motor behaviors are initiated, such as chewing, licking, and swallowing. These motor behaviors are organized at the brainstem level and can occur even after decerebration. Another reflex involves the facial expressions that occur in response to taste stimuli. Both humans and rats make stereotypical facial expressions to taste qualities and these have been extensively studied and used as a behavioral measure of taste hedonics.[5] In humans, these expressions can be elicited at birth before any experience of taste stimuli has occurred and are, therefore, considered to be innate, implying developmentally determined rNST circuits.[6]

Moreover, this reflex behavior is retained in decerebrate animals, again indicating innate brainstem circuits.[7] Finally, saliva is secreted in response to taste and somatosensory oral stimulation. Because salivary secretion is solely initiated in response to activation of the autonomic nerve supply to the salivary glands,[8] sensory information from the oral cavity must form the afferent limb of this reflex. The gustatory component of these essential reflexes, therefore, is processed and initiated by the rNST and reflects complex neural mechanisms that occur when sensory information arrives at the nucleus.

4.2 NEUROMUSCULAR-RELATED REFLEX ACTIVITY

As early as 1909, Ramón Y Cajal[9] described the neuromuscular reflex activity initiated by feeding. He states

"Imagine that food acts as a stimulus on lingual and glossopharyngeal nerve endings. As indicated in Figure 4.1 [see also Figure 1 in Miller,[3]], impulses course along the two nerves to their respective ganglia and then along the central branches to the corresponding medullary sensory nuclei, where the two nerves give rise to abundant collaterals. Within these nuclei, impulses pass from the collaterals to cells that are contacted by the collaterals, and then to the axon of cells in the sensory nuclei. As is well known, these axons form the central (second order) pathways associated with the trigeminal and glossopharyngeal nerves. Collaterals arising from these pathways distribute impulses to hypoglossal neurons, and the tongue muscles contract. It is quite likely that this circuit mediates the various reflexes activated by inputs from the lingual nerve, including *chewing, swallowing, sucking,* and so on." (Translated by Swanson and Swanson.[10])

Thus, Cajal had already defined the basic reflex circuit from sensory endings in the dorsal tongue mucosa to the hypoglossal motor nucleus. Note that the sensory contribution of the facial nerve input is missing from Cajal's diagram, because at the time he published his description, there was still some confusion regarding the contribution of the sensory part of the facial nerve to the solitary tract and nucleus (see Chapter 1). Cajal also failed to detail the contribution of gustatory and somatosensory input to the hypoglossal reflex. Functional details of the reflex circuits described by Cajal inspired a number of later investigators.

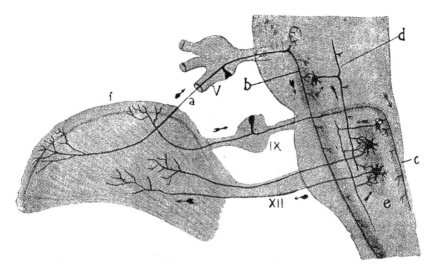

FIGURE 4.1 Diagram by Cajal showing the direction of current flow through the trigeminal, glossopharyngeal, and vagus nerves. The glossopharyngeal nerve (IX) is shown forming the descending solitary tract (c) and as indicated by the arrow conducts information to the brainstem. The hypoglossal nucleus (e) conducts information from the brainstem via the hypoglossal nerve (XII) to the tongue muscles. a, lingual nerve; V, trigeminal ganglion; b, spinal nucleus of the trigeminal nerve; d, central pathway of the trigeminal system; f, lingual mucosa. (From Ramón Y. Cajal, S. *Histologie du Système Nerveux de l'Homme et des Vertébrés*, 1909, p. 579, with permission.)

4.2.1 PHYSIOLOGICAL INVESTIGATIONS OF GUSTO-LINGUAL REFLEXES

The physiological study of reflex orofacial activity was pioneered by Sherrington,[11,12] who noted in decerebrate cat preparations that mechanical stimulation of the tongue filiform papillae resulted in tongue movements. Electrical stimulation of the combined chorda-lingual nerve, which contains the afferent fibers innervating both taste and somatosensory receptors, also resulted in reflex tongue movement. Thus, the afferent limb described in Cajal's anatomical descriptions of reflex tongue movements was established. Electrical stimulation of the chorda-lingual nerve evokes neural discharges in the hypoglossal nerve with a latency of 7 ms as well as synaptic potentials in the hypoglossal motoneurons.[13,14] Based on this latency, only two synapses are involved in the reflex, and one of these is between the afferent fibers and either the NST or trigeminal sensory nuclei, leading to the conclusion that only one internuncial relay is involved in this reflex.[15] Synaptic potentials and reflex discharges in the hypoglossal motoneurons were also elicited by mechanical stimulation of the dorsal tongue surface.[15,16] A similar tongue reflex can be elicited by stimulation of the glossopharyngeal nerve with a slightly longer latency.[17]

In most of these earlier studies, either tactile or nociceptive stimulation is assumed to be the sensory system responsible for the afferent limb of the reflex, although Porter[14] entertained the possibility that chorda tympani fibers might be

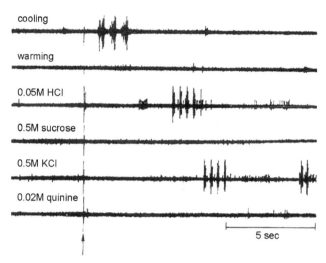

FIGURE 4.2 Electrophysiological responses evoked in the rat hypoglossal nerve to taste and thermal stimuli applied (indicated by the arrow) to the tongue. (Reprinted from Yamamoto, T., *Brain Res.*, 92, 499–504, 1975, with permission from Elsevier.)

activated by the lingual nerve stimulation. Kawamura[18] suggested that other sensory modalities besides tactile stimulation may also be important in feedback regulation of tongue movements. This suggestion was subsequently tested by Yamamoto[19] using mechanical, gustatory, and thermal stimuli to demonstrate that application of KCl, NaCl, HCl, and cooling stimuli to the tongue all evoked efferent neural activity in the hypoglossal nerve (Figure 4.2). However, these taste-initiated reflex activities in the hypoglossal nerve had very long latencies (mean = 6 sec) that do not compare to the behavioral latencies, throwing some doubt on the significance of the electrophysiological experiments. In behavioral experiments, Weiffenbach and Thach[20] and Nowlis[21] reported that drops of glucose solution applied to the tongue tip in infants elicited lateral tongue movements after a short latency (Weiffenbach, personal communication). Moreover, in both the animal and human studies, these tongue reflex responses could be elicited without the involvement of rostral brain centers.[6]

4.2.2 Physiological Investigations of Gusto-Facial Reflexes

Little information is available on the functional connections between afferent taste input and motoneurons in the facial nucleus responsible for the stereotypical responses to taste stimuli. Most of the projections to the facial motor nucleus originate from neurons in the reticular formation[22] but some derive from the trigeminal complex,[23,24] and even fewer connections are reported projecting from the NST.[25,26] However, experiments with lesioned decerebrate animals demonstrate that the circuits responsible for mimetic responses to taste stimuli are at the brainstem level.[7] The results of lesions in NST and PBN indicate marked changes in the taste reactivity test elicited by taste stimuli.[27] Facial nerve responses to tongue stimulation with taste stimuli have been reported, and neural activity elicited by citric acid and

NaCl occurs after a short latency.[6] However, these experiments were very preliminary and merit further analysis in additional animals.

4.3 GUSTO-SALIVARY REFLEX ACTIVITY

A key component of the oral milieu is saliva secreted by the major and minor salivary glands. Saliva, although mainly consisting of water, has numerous functions,[28,29] such as providing the lubrication necessary for movements of speech, mastication, and swallowing. It provides a buffer to stabilize intraoral pH. An indication of the importance of saliva for oral health is demonstrated when it is absent or reduced, resulting in a fetid mouth due to intraoral bacterial proliferation. Beside these numerous vital functions, saliva is essential in the initial stages of taste transduction, because it acts as a solvent for taste stimuli, a transport medium for the dissolved stimuli, and a possible source of ions that pass through taste receptor apical ion channels to depolarize or hyperpolarize taste receptor cells.

4.3.1 TASTE-INITIATED REFLEX SECRETION OF SALIVA FROM THE MAJOR SALIVARY GLANDS

To provide the saliva necessary for these functions, stimulation of taste receptors reflexly initiates saliva flow.[30,31] The reflex does not merely switch the flow of saliva on and off, but is influenced by the sensory properties of the stimulus. For example, citric acid stimulation initiates high salivary flow rates rich in bicarbonate acting as a buffer, whereas sweet-tasting stimuli result in a lower flow rate of saliva containing salivary amylase.[32–35] Moreover, this modality-specific reflex activity occurs in decerebrate animals.[36] Most of these results were obtained in anesthetized animals; in experiments investigating reflex salivary secretion in awake, behaving animals, some interesting differences were reported.[37] For example, recordings of salivary flow and licking behavior reveal that salivary flow is greatest during grooming behavior (Figure 4.3A) and while eating pellet food (Figure 4.3B) and is considerably less with sweet, sour, and salty stimuli (Figure 4.3C). Interestingly, salivary flow to taste stimuli was highest after quinine infusion into the oral cavity (Figure 4.3Da and 4.3Db), presumably in an effort to remove the aversive stimulus from the mouth. However, because the animals used in these studies are awake, forebrain centers involved in the control of feeding may also contribute to the control of the salivary nucleus neurons in the brainstem.

4.3.2 TASTE-INITIATED REFLEX SECRETION OF SALIVA FROM VON EBNER'S SALIVARY GLANDS

Although salivary secretion from the major salivary glands is important to provide saliva bathing the oral mucosa, taste buds in the circumvallate and foliate papillae are situated in clefts in the tongue surface. Von Ebner's lingual salivary glands drain into the clefts of these papillae and are responsible for providing the microenvironment of the taste buds. Taste stimuli reach the circumvallate and foliate taste buds via the diffusion pathway provided by von Ebner's gland secretions, and the glands

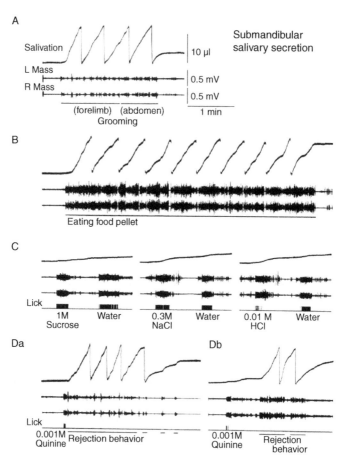

FIGURE 4.3 Recordings of submandibular salivary secretion, jaw movement, and licking from an unanesthetized rat. Each tracing consists of salivary flow rate (Salivation), electo-myographic activity of the left and right masseter muscles (L Mass, R Mass), and licking (Lick in records C and D). Observed behavior is indicated by a line underneath the traces. (A) Salivary secretion evoked by grooming. (B) Salivary secretion evoked by eating a food pellet. (C) Salivation in response to taste stimuli. Taste stimuli do not result in high flow rates of saliva. (Da and Db) Salivation in response to licking quinine, which is only briefly sampled but evokes prolonged salivation and mouth movements. (From Matsuo, R. et al., *Brain Res.*, 649, 138, 1994. With permission.)

are also responsible for flushing out taste stimuli as well as maintaining a healthy cleft environment. Thus, there is cooperative interaction between the taste buds in the circumvallate and foliate papillae and von Ebner's gland secretions. This inter-action was examined by taking advantage of the fact that the single centrally placed circumvallate papilla in the rat is bilaterally innervated. Recordings of taste stimu-lation of the circumvallate papilla could be made from one glossopharyngeal nerve while saliva flow to the papilla cleft was initiated by electrical stimulation of the contralateral glossopharyngeal nerve.[38] Salivary secretion from von Ebner's glands

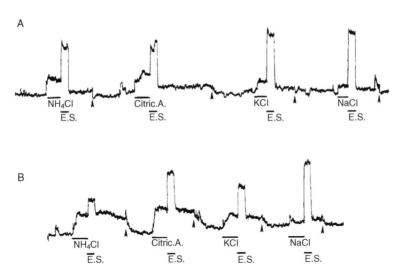

FIGURE 4.4 (A) Summated whole-nerve recordings of ipsilateral glossopharyngeal nerve responses to 0.5 M NH$_4$Cl, citric acid, KCl, and NaCl applied to the circumvallate papilla in a rat. During the recording of the taste response, the contralateral glossopharyngeal nerve was briefly electrically stimulated (E.S.) to evoke salivary secretion from von Ebner's glands that drain into the clefts of the circumvallate papilla. The steady-state taste response is reduced by approximately 50% for all stimuli. Water is flowed over the tongue (arrowhead) to rinse the taste stimulus away. (B) Similar experiment as in A, after atropine had been systemically administered to the animal to block salivary secretion. The electrical stimulation has no effect on the taste responses. Calibration bar = 30 sec. (From Gurkan, S. and Bradley, R. M., *Chem. Senses*, 13, 655, 1988, by permission of Oxford University Press.)

significantly reduced taste responses to stimulation of the circumvallate papilla by taste stimuli (Figure 4.4). Because the reduction in taste responses was blocked by the parasympathetic antagonist atropine, which acts at the gland synapses, the reduction in taste response was concluded to be due to saliva flow washing the stimulus from the receptors.

4.3.3 BRAINSTEM SALIVATORY NUCLEI

The gustatory-salivary reflexes involve afferent input to the rNST that then synapses with parasympathetic secretomotor neurons innervating the salivary glands. The parasympathetic preganglionic neurons controlling the salivary glands are collected in a column of brainstem neurons termed the *salivatory nuclei*. Efferent axons of the preganglionic neurons synapse with cells in peripheral autonomic ganglia that then send axons to synapse with the secretory cells of the salivary glands.[39] The parasympathetic neurons integrate input from both the rNST and forebrain areas and form the final common pathway to the salivary glands.

Numerous pathway tracing and electrical stimulation studies have defined the efferent connections from the brainstem to the salivary glands.[40] The salivatory nuclei are divided into a superior and inferior division based on the association of their preganglionic axons with a cranial nerve. There is no anatomical demarcation

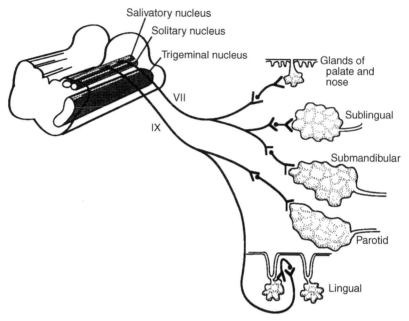

FIGURE 4.5 Schematic of the parasympathetic innervation of the salivary glands. The cell bodies of the secretomotor neurons lie in the salivary nucleus situated adjacent and medio-ventral to the solitary nucleus. Efferent axons from the salivary neurons travel in the facial (VII) and glossopharyngeal (IX) nerves. Salivary neurons with axons in the facial nerve are situated in the superior salivary nucleus, whereas salivary neurons with axons in the glossopharyngeal nerve are situated in the inferior salivary nucleus. The axons from the salivary neurons (preganglionic) synapse in a peripheral ganglion, which gives rise to postganglionic fibers that synapse on the salivary gland acinar cells. The inferior salivary nucleus innervates the parotid and von Ebner's (lingual) salivary glands. The superior salivary nucleus controls the sublingual and submandibular salivary glands.

between the superior and inferior divisions of the salivary nuclei. Cell bodies of preganglionic fibers innervating the submandibular and sublingual salivary glands form the superior salivary nucleus (SSN) with efferent fibers that travel with the chorda tympani nerve (VII) to the glands. Cell bodies of preganglionic fibers innervating the parotid and von Ebner's glands form the inferior salivary nucleus (ISN) with efferent fibers that travel with the glossopharyngeal nerve (IX) to the glands (Figure 4.5).

4.3.4 NEUROBIOLOGY OF THE SALIVATORY NEURONS

Neurons of the ISN are situated along the medial border of the rNST (Figure 4.6A). They have been subjected to detailed morphological analysis and found to vary in the complexity of their dendritic trees.[41] Neurons of the ISN innervating the parotid and von Ebner's glands consist of two separate cell groups; the most ventral in the nucleus supplies von Ebner's glands, whereas the more dorsal

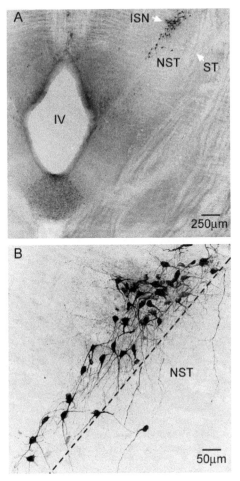

FIGURE 4.6 (A) Horizontal section through the rat brainstem showing the nucleus of the solitary tract (NST) and retrogradely labeled neurons of the inferior salivatory nucleus (ISN). Rostral is towards the top of the figure. IV, fourth ventricle; ST, solitary tract. Merged confocal image, × 100. (B) Higher power view of the neurons of the inferior salivatory nucleus. The medial border of the nucleus of the solitary tract (NST) is indicated by the dashed line. Merged confocal image, × 400. (Reprinted from Kim, M., Chiego, D. J. and Bradley, R. M., *Auton. Neurosci.*, 111, 27–36, 2004, with permission from Elsevier.)

neurons innervate the parotid gland (Figure 4.7). The parotid preganglionic neurons are also larger than the neurons innervating von Ebner's glands (Figure 4.8). Dendrites of the salivatory neurons penetrate into the rNST, facilitating synaptic contacts, and a few ISN neurons are actually situated within the rNST (Figure 4.6B). By double-labeling the afferent input to the rNST and the neurons of the ISN, afferent fibers are seen to travel to the ISN neurons, suggesting monosynaptic connections between the afferent taste fibers and the parasympathetic efferent neuron (Figure 4.9)

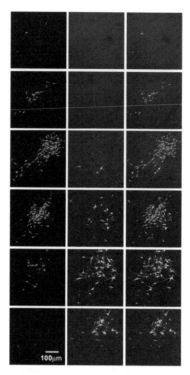

FIGURE 4.7 (See color insert following p.78) Confocal images of inferior salivatory nucleus neurons (ISN). Neurons innervating the parotid gland were labeled with a green fluorescent tracer, and neurons innervating von Ebner's glands were labeled with a red fluorescent tracer. The left column is ISN neurons innervating the parotid gland, and the center column is the ISN neurons innervating von Ebner's glands. The right column is the merged images of the left and center columns. Each row is a merged image of a 100 μm section. The top row is the most dorsal section and the bottom row the most ventral. Note that neurons innervating the parotid and von Ebner's glands are separate populations and that the neurons innervating von Ebner's glands are situated more dorsal than the neurons innervating the parotid gland.

Attempts have been made to study synaptic connections between the afferent input and the salivatory nucleus neurons. Because the parasympathetic fibers travel in the chorda tympani and glossopharyngeal nerves, salivatory neurons can be electrophysiologically identified by antidromic electrical stimulation.[42–44] It was found that electrical stimulation resulted in the generation of a few action potentials with variable latency ranging from 10 to 85 ms, suggesting that the gustatory reflex involves multiple synapses between the afferent input and the efferent output. However, it was not possible to confirm whether the recordings actually originated from salivatory neurons using this methodology. More recently, by prelabeling ISN and SSN neurons, it has become possible to identify them in brain slices and make whole-cell recordings from the neurons.[45,46] Recordings from identified neurons of the salivatory nuclei have demonstrated that the postsynaptic potentials initiated by stimulation of the solitary tract or adjacent neuropil are a mixture of excitation and inhibition (Figure 4.10).[47,48] Applications of synaptic agonists and antagonists have revealed that

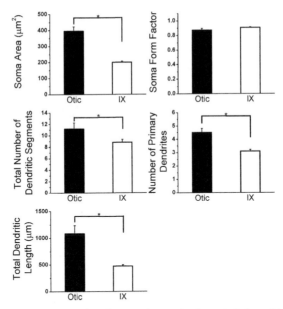

FIGURE 4.8 Histograms comparing the morphometric characteristics of inferior salivatory neurons innervating the parotid and von Ebner's salivary glands. The parotid gland neurons were labeled via the otic ganglion (Otic) and von Ebner's gland neurons were labeled by application of the tracer to the lingual branch of the glossopharyngeal nerve (IX). The soma area, total number of dendritic segments, total dendritic length, and number of primary dendrites were all significantly larger for the parotid gland neurons as compared to von Ebner's gland neurons. The soma form factor (a measure of the roundness of the soma shape) was similar for neurons innervating both glands.

the excitatory component is mediated by NMDA and AMPA glutamate receptors and that inhibition results from stimulation of $GABA_A$ receptors.[47,49] Glycinergic receptors have been demonstrated on some salivatory neurons, also.[47] Somata and dendrites of salivatory neurons receive symmetric synapses from glutamatergic, GABA, and glycine immunoreactive varicosities.[50] Salivatory nucleus neurons were shown to express NMDA, AMPA, and kainate glutamate receptor subtypes. Labeling by NMDA receptor subtypes was especially prominent. NR2B stained the nuclei strongly, whereas NR1, NR2A, and GluR1 receptor subtypes were also expressed in the dendrites, as well as the neuron cell bodies.[51] Based on the neurophysiological measures, these excitatory and inhibitory synapses either are monosynaptic connections between primary afferent fibers or involve polysynaptic connections via neurons in the rNST.[49]

The salivary reflex is also influenced by descending input from higher brain centers,[30] and a number of neuropeptides have been hypothesized to be involved in this descending modulation of secretion as demonstrated by strong immunoreactivity for substance P, serotonin (5HT), neuropeptide Y, somatostatin, tyrosine hydroxylase, vasoactive intestinal peptide, and calcitonin gene-related peptide.[52] Of these neuropeptides, substance P and 5HT have been shown to excite salivatory neurons in a concentration-dependent manner.[53,54]

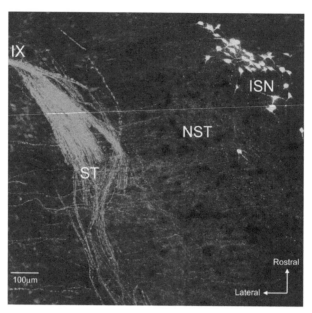

FIGURE 4.9 (See color insert following p.78) Merged confocal image of the contribution of the glossopharyngeal nerve to the solitary tract (red, ST) and ISN neurons innervating von Ebner's gland (green). The fibers of the solitary tract give off collateral branches, some of which reach the ISN neurons, suggesting that some afferent fibers make monosynaptic contacts with the ISN neurons. Other collaterals synapse with neurons of the NST, which are unstained but appear as dark voids among the terminal field of the glossopharyngeal nerve. This image represents the basic circuit of the gusto-salivary reflex.

4.4 FUTURE DIRECTIONS

Neurobiologists have long used reflex activity to probe the function of neural circuits, and because there are several well-defined reflexes that involve the NST, it will be possible to use these reflexes to define how the NST functions. In particular, gusto-salivary reflexes seem ideal for this purpose. The afferent limb consists of input from somatosensory and gustatory receptors, and the motor output is made up of parasympathetic secretomotor neurons. Using brain slices, the afferent input can be stimulated, and by prelabeling the parasympathetic neurons, the efferent limb of the reflex arc can be visualized for recording. Indeed, this has already been utilized by the Bradley group as well as Matsuo's group in Japan. The task ahead is to define the role of the rNST neurons in this reflex. Because various taste modalities are known to initiate different types of saliva, it will be important to determine the underlying connections responsible for these differences. For example, because acid stimuli initiate high flow rates of saliva concentrated with bicarbonate buffer, is it possible that acid-best fibers make direct monosynaptic contacts with the parasympathetic neurons? Are the afferent inputs segregated so that some connect to rostral directed taste pathways, while others are connected to salivatory nucleus neurons? What are the roles of descending connections to the salivatory nucleus neurons and how do they interact with the

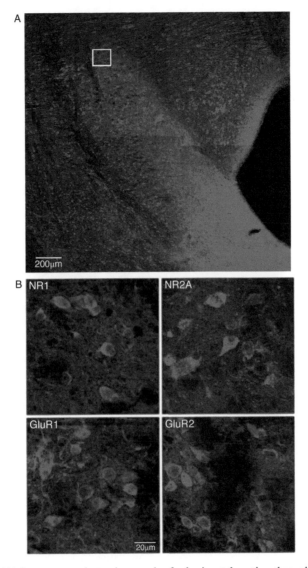

FIGURE 3.4 (A) Low-power photomicrograph of a horizontal section through the left half of the rat medulla. The 4ᵗʰ ventricle is on the right and rostral is towards the top of the micrograph. The section has been immunoreacted with antibodies to the NR2 glutamate receptor subtype. The NST immunoreacts strongly for the NR2 receptor subtype. The solitary tract is unstained. The small white square indicates the location of the higher power photomicrographs in B. (B) Immunoreactions to the NR1, NR2A, GluR1, and GluR2 glutamate receptor subtypes. Neurons in the NST express all four of these glutamate receptor subtypes.

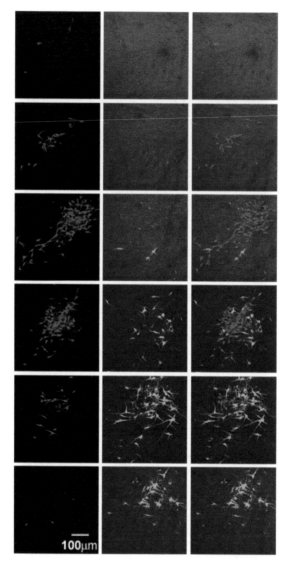

FIGURE 4.7 Confocal images of inferior salivatory nucleus neurons (ISN). Neurons innervating the parotid gland were labeled with a green fluorescent tracer, and neurons innervating the von Ebner glands were labeled with a red fluorescent tracer. The left column is the ISN neurons innervating the parotid gland, and the center column is the ISN neurons innervating the von Ebner glands. The right column is the merged images of the left and center columns. Each row is a merged image of a 100 μm section. The top row is the most dorsal section and the bottom row the most ventral. Note that neurons innervating the parotid and von Ebner glands are separate populations and that the neurons innervating the von Ebner glands are situated more dorsal than the neurons innervating the parotid gland.

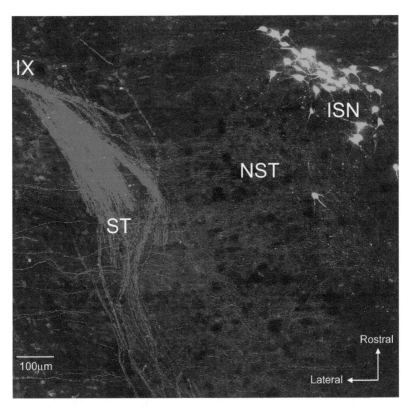

FIGURE 4.9. Merged confocal image of the contribution of the glossopharyngeal nerve to the solitary tract (red, ST) and ISN neurons innervating the von Ebner gland (green). The fibers of the solitary tract give off collateral branches, some of which reach the ISN neurons, suggesting that some afferent fibers make monosynaptic contacts with the ISN neurons. Other collaterals synapse with neurons of the NST, which are unstained but appear as dark voids among the terminal field of the glossopharyngeal nerve. This image represents the basic circuit of the gusto-salivary reflex.

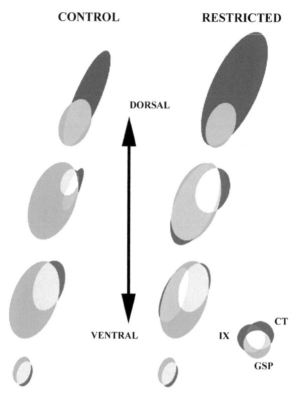

CONTROL **RESTRICTED**

DORSAL

VENTRAL IX CT

GSP

FIGURE 6.11 Model of terminal field organization through the dorsal-ventral extent of horizontal sections from the right NST in control and restricted rats. Overlapping fields are represented at four levels along the dorsal-ventral axis. Refer to key to identify individual fields and respective overlap among different terminal fields. Note: Due to the orientation of the NST within the brainstem, the term "dorsal" sections refers to dorsal-caudal and "ventral" refers to ventral-rostral. (From May, O. L. and Hill, D. L., *J. Comp. Neurol.,* 497, 658, 2006, with permission.)

A

a Total EPSCs

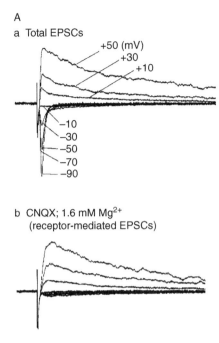

b CNQX; 1.6 mM Mg^{2+}
  (receptor-mediated EPSCs)

c CPP+CNQX

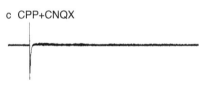

FIGURE 4.10 Excitatory postsynaptic currents (EPSC) recorded from neurons of the superior salivatory nucleus. Synaptic currents were evoked by electrical stimulation close to the neuron. (A)(a) Total current recorded at potentials between –90 and +50 mV voltage steps. (A)(b) EPSC recorded after application of the AMPA/kianate receptor blocker CNQX is reduced in amplitude. (A)(c) The EPSC was totally eliminated after application of both CNQX and an NMDA receptor blocker CPP. These results indicate that both NMDA and AMPA/kianate receptors are involved in excitatory synaptic activity in salivatory neurons. (Reprinted from Mitoh, Y., Funahashi, M., Kobashi, M., and. Matsuo, R., *Brain Res.*, 999, 62, 2004, with permission from Elsevier.)

ascending input? Much recent experimental evidence has indicated that the rNST is not a simple relay nucleus, and study of these reflexes will contribute much new knowledge to our understanding of how complex the rNST really is.

ACKNOWLEDGMENT

The preparation of this chapter was supported in part by NIH grant DC 000288 from the National Institute on Deafness and Other Communication Disorders to R. M. Bradley.

REFERENCES

1. Andresen, M. C. and Kunze, D. L. Nucleus tractus solitarius—Gateway to neural circulatory control, *Annu. Rev. Physiol.*, 56, 93, 1994.
2. Saper, C. B. Central autonomic system, in *The Rat Nervous System,* Paxinos, G. (Ed.), Elsevier Academic Press, Amsterdam, 2004, 761.
3. Miller, A. J. Oral and pharyngeal reflexes in the mammalian nervous system: their diverse range of complexity and the pivotal role of the tongue, *Crit. Rev. Oral Bio. Med.*, 13, 409, 2002.
4. Lowe, A. A. The neural regulation of tongue movements, *Prog. Neurobiol.*, 15, 295, 1981.
5. Grill, H. J. and Norgren, R. The taste reactivity test. I. Mimetic responses to gustatory stimuli in neurologically normal rats, *Brain Res.*, 143, 263, 1978.
6. Steiner, J. E. The gustofacial response: observation on normal and anencephalic newborn infants, in *Fourth Symposium on Oral Sensation and Perception. Development in the Fetus and Infant,* Bosma, J. F. (Ed.), Department Health, Education, and Welfare, Bethesda, MD, 1973, 254.
7. Grill, H. J. and Norgren, R. The taste reactivity test. II. Mimetic responses to gustatory stimuli in chronic thalamic and chronic decerebrate rats, *Brain Res.*, 143, 281, 1978.
8. Schneyer, L. H., Young, J. A. and Schneyer, C. A. Salivary secretion of electrolytes, *Physiol. Rev.*, 52, 720, 1972.
9. Ramón Y Cajal, S. *Histologie du Système Nerveux de l'Homme et des Vertébrés.* Maloine, Paris, 1909.
10. Ramón Y Cajal, S. *Histology of the Nervous System.* Swanson, N. and Swanson, L. W. (Eds.), Oxford University Press, New York, 1995.
11. Miller, F. R. and Sherrington, C. S. Some observations on the buccopharyngeal stage of reflex deglutition in the cat, *Quart. J. Exp. Physiol.*, 9, 147, 1916.
12. Sherrington, C. S. Reflexes elicitable in the cat from pinna vibrissae and jaws, *J. Physiol. (Lond.)*, 51, 431, 1917.
13. Blom, S. and Skoglund, S. Some observations on the control of tongue muscles, *Experientia*, 15, 12, 1959.
14. Porter, R. Synaptic potentials in hypoglossal motoneurones, *J. Physiol. (Lond.)*, 180, 209, 1965.
15. Porter, R. The synaptic basis of a bilateral lingual-hypoglossal reflex in cats, *J. Physiol. (Lond.)*, 190, 611, 1967.
16. Porter, R. Synaptic influences of lingual exteroceptors on hypoglossal motoneurones. *J. Physiol.*, 184, 24P, 1966.
17. Blom, S. Afferent influences on tongue muscle activity. *Acta Physiol. Scand.* 49 (Supplementum 170), 1, 1960.
18. Kawamura, Y. Neuromuscular mechanisms of jaw and tongue movement, *J. Amer. Dent. Assoc.*, 62, 545, 1961.
19. Yamamoto, T. Linguo-hypoglossal reflex: effects of mechanical thermal and taste stimuli, *Brain Res.*, 92, 499, 1975.
20. Weiffenbach, J. M. and Thach, B. T. Elicited tongue movements: touch and taste in the mouth of the neonate, in *Fourth Symposium on Oral Sensation and Perception. Development in the Fetus and Infant,* Bosma, J. F. (Ed.), Department of Health, Education, and Welfare, Bethesda, MD, 1973, 232.
21. Nowlis, G. H. Taste-elicited tongue movement in human newborn infants: an approach to palatability, in *Fourth Symposium on Oral Sensation and Perception,* Bosma, J. F. (Ed.), Department of Health, Education, and Welfare, Bethesda, MD, 1973, p. 292.

22. Travers, J. B. Efferent projections from the anterior nucleus of the solitary tract of the hamster, *Brain Res.*, 457, 1, 1988.

23. Erzurumlu, R. S. and Killackey, H. P. Efferent connections of the brainstem trigeminal complex with the facial nucleus of the rat, *J. Comp. Neurol.*, 188, 75, 1979.

24. Pinganaud, G., Bernat, I., Buisseret, P. and Buisseret-Delmas, C. Trigeminal projections to hypoglossal and facial motor nuclei in the rat, *J. Comp. Neurol.*, 415, 91, 1999.

25. Norgren, R. Projections from the nucleus of the solitary tract in the rat, *Neuroscience*, 3, 207, 1978.

26. Travers, J. B. and Norgren, R. Afferent projections to the oral motor nuclei in the rat, *J. Comp. Neurol.*, 220, 280, 1983.

27. Flynn, F. W., Grill, H. J., Schwartz, G. J. and Norgren, R. Central gustatory lesions: I. Preference and taste reactivity tests, *Behav. Neurosci.*, 105, 933, 1991.

28. Bradley, R. M. Salivary secretion, in *Smell and Taste in Health and Disease,* Getchell, T. V. et al. (Eds.), Raven Press, New York, 1991, 127.

29. Bradley, R. M. and Beidler, L. M. Saliva: its role in taste function, in *Handbook of Olfaction and Gustation,* Doty, R. L. (Ed.), Marcel Dekker, New York, 2003, 639.

30. Matsuo, R. Interrelation of taste and saliva, in *Neural Mechanisms of Salivary Gland Secretion,* Garrett, J. R. et al. (Eds.), Karger, Basel, Switzerland, 1999, 185.

31. Hector, M. P. and Linden, R. W. A. Reflexes of salivary secretion, in *Neural Mechanisms of Salivary Gland Secretion,* Garrett, J. R. et al. (Eds.), Karger, Basel, Switzerland, 1999, 196.

32. Kawamura, Y. and Yamamoto, T. Studies on neural mechanisms of the gustatory-salivary reflex in rabbits, *J. Physiol.*, 285, 35, 1978.

33. Kemmer, T. and Malfertheiner, P. Der differenzierte Einflus der Geschmacksqualitaten "sub" und "sauer" auf die Parotissekretion, *Res. Exp. Med.*, 183, 35, 1983.

34. Newbrun, E. Observations on the amylase content and flow rate of human saliva following gustatory stimulation, *J. Dent. Res.*, 41, 459, 1962.

35. Gjörstrup, P. Taste and chewing as stimuli for the secretion of amylase from the parotid gland of the rabbit, *Acta Physiol. Scand.*, 110, 295, 1980.

36. Matsuo, R., Yamamoto, T., Yoshitaka, K. and Morimoto, T. Neural substrates for reflex salivation induced by taste, mechanical, and thermal stimulation of the oral region in decerebrate rats. *Jpn. J. Physiol.* 39, 349, 1989.

37. Matsuo, R., Yamamoto, T., Ikehara, A. and Nakamura, O. Effect of salivation on neural taste responses in freely moving rats: analyses of salivary secretion and taste responses of the chorda tympani nerve, *Brain Res.*, 649, 136, 1994.

38. Gurkan, S. and Bradley, R. M. Secretions of von Ebner's glands influence responses from taste buds in rat circumvallate papilla, *Chem. Senses*, 13, 655, 1988.

39. Loewy, A. D. Anatomy of the autonomic nervous system: an overview, in *Central Regulation of Autonomic Function,* Loewy, A. D. and Spyer, K. M. (Eds.), Oxford University Press, New York, 1990, 3.

40. Matsuo, R. Central connections for salivary innervation and efferent impulse formation, in *Neural Mechanisms of Salivary Secretion,* Garrett, J. R. et al. (Eds.), Karger, Basel, Switzerland, 1999, 26.

41. Kim, M., Chiego, D. J. and Bradley, R. M. Morphology of parasympathetic neurons innervating the lingual salivary glands, *Auton. Neurosci.*, 111, 27, 2004.

42. Eisenman, J. S. Response of rat superior salivatory units to chorda tympani stimulation, *Brain Res. Bull.*, 10, 811, 1983.

43. Murakami, T., Ishizuka, K. and Uchiyama, M. Convergence of excitatory inputs from the chorda tympani, glossopharyngeal and vagus nerves onto superior salivatory nucleus neurons in the cat, *Neurosci. Lett.*, 105, 96, 1989.

44. Ishizuka, K. and Murakami, T. Convergence of excitatory inputs from the chorda tympani, glossopharyngeal, and vagus nerves onto inferior salivatory nucleus neurons in the cat, *Neurosci. Lett.*, 143, 155, 1992.

45. Matsuo, R. and Kang, Y. Two types of parasympathetic preganglionic neurons in the superior salivatory nucleus characterized electrophysiologically in slice preparations of neonatal rats, *J. Physiol. (Lond.)*, 513, 157, 1998.

46. Fukami, H. and Bradley, R. M. Biophysical and morphological properties of parasympathetic neurons controlling the parotid and von Ebner salivary glands in rats, *J. Neurophysiol.*, 93, 678, 2005.

47. Mitoh, Y., Funahashi, M., Kobashi, M. and Matsuo, R. Excitatory and inhibitory postsynaptic currents of the superior salivatory nucleus innervating the salivary glands and tongue in the rat, *Brain Res.*, 999, 62, 2004.

48. Bradley, R. M., Fukami, H. and Suwabe, T. Neurobiology of the gustatory-salivary reflex, *Chem. Senses*, 30, i70–i71, 2005.

49. Suwabe, T. and Bradley, R. M., Characterization of the gustosalivary reflex: excitatory postsynaptic potentials recorded from identified inferior salivatory nucleus neurons, Abstract Viewer/Intinerary Planner, Society for Neuroscience, Washington, DC, 281.3, 2005.

50. Kobayashi, M., Nemoto, T., Nagata, H., Konno, A. and Chiba, T. Immunohistochemical studies on glutamatergic, GABAergic and glycinergic axon viscosities presynaptic to parasympathetic preganglionic neurons in the superior salivatory nucleus of the rat, *Brain Res.*, 766, 72, 1997.

51. Kim, M., Chiego, D. J., and Bradley, R. M., Characterization of the gusto-salivary reflex: ionotopic glutamate receptor expression in preganglionic neurons of the rat inferior salivatory nucleus, in *Abstract Viewer/Itinerary Planner,* Society for Neuroscience, Washington, DC, 281.4, 2005.

52. Nemoto, T., Konno, A. and Chiba, T. Synaptic contact of neuropeptide- and amine-containing axons on parasympathetic preganglionic neurons in the superior salivatory nucleus of the rat, *Brain Res.*, 685, 33, 1995.

53. Suwabe, T. and Bradley, R. M. Effect of substance P on parasympathetic neurons controlling the lingual salivary glands, Abstract Viewer/Intinerary Planner, Society for Neuroscience, Washington, DC, 179.1, 2004.

54. Suwabe, T. and Bradley, R. M. Effect of serotonin on membrane properties of neurons of the rat inferior salivatory nucleus, *Chem. Senses,* 30, 472, 2005.

5 Neural Coding in the rNST

David V. Smith and Christian H. Lemon

CONTENTS

5.1   INTRODUCTION

The gustatory system in mammals provides sensory input that is critical for the regulation of ingestive behavior and the avoidance of toxic substances. Taste serves a unique role among sensory systems in the extent to which it interfaces with neural substrates of reward and motivation.[1] For example, sweet- and bitter-tasting stimuli produce inherent preference and avoidance, respectively, and sweet-tasting stimuli can serve as effective reinforcers. The anatomy of the taste system reflects its dual role as both a discriminative system, designed to determine subtle differences in taste quality and intensity, and a motivational one, which underlies the acceptance and rejection of potential foods. Anatomically, the taste system is situated between the external environment and the internal milieu, making taste a rostral extension of the visceral afferent system.[2]

The rostral nucleus of the solitary tract (rNST) in the medulla lies at the interface between taste receptors and the brain. Gustatory neurons in the rNST receive input from the chorda tympani and greater superficial petrosal nerves, which are branches of cranial nerve VII, the glossopharyngeal nerve (IX), and the vagus nerve (X). The chorda tympani nerve innervates taste bud cells of the fungiform papillae on the anterior tongue, whereas the greater superficial petrosal nerve innervates those in palatal epithelia. The glossopharyngeal nerve relays taste information from the posterior tongue to the rNST. The vagus nerve innervates receptor cells found near the epiglottis. Gustatory neurons in the rNST serve the critical function of integrating and encoding information received from taste receptors and routing this information to higher centers involved with motivated behavior and perceptual processing. The NST also projects information to nearby structures in the brainstem that are involved in mediating oromotor responses. Many factors influence how rNST neurons respond to taste stimuli and also how information about tastants could be encoded in their spike outputs.

5.2 TASTE RESPONSES OF rNST NEURONS

Gustatory neurons in the rNST are typically more broadly tuned to taste stimuli than peripheral fibers and often respond to somatosensory inputs, such as tactile and temperature stimulation, as well.[3,4] Neurons in the rNST receive converging input from peripheral gustatory axons,[5] and these inputs can arise from separate taste bud populations, such as those on the anterior tongue and palate.[6,7] This convergence renders rNST neurons more broadly tuned than peripheral fibers to the extent that medullary taste cells are quite broadly responsive.[3] The responses of an NST neuron of the rat to several taste stimuli are shown in Figure 5.1. As predicted by the co-expression of T2r taste receptors for bitter stimuli in taste bud cells,[8] the responses to bitter stimuli in the cell in Figure 5.1 and other bitter-sensitive NST cells are highly correlated,[9] although these same neurons respond to stimuli representing other taste qualities as well, such as NaCl and HCl (Figure 5.1).

5.2.1 Breadth of Tuning

The breadth of tuning of sensory neurons is an important parameter that impacts the information-handling characteristics of such cells. In taste, breadth of tuning has been described by the number of stimuli among those representing the basic taste qualities (i.e., sweet, salty, sour, and bitter) that produce responses in a neuron.[10,11] However, this method depends heavily on the experimenter's definition of a taste response. In 1979, Smith and Travers[12] introduced the entropy measure as a way to

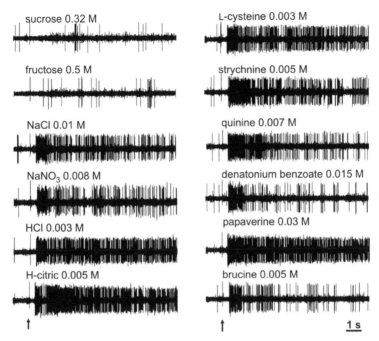

FIGURE 5.1 Response properties of an individual gustatory neuron in the rat NST. Oscilloscope sweeps show single-unit activity evoked by the application of several stimuli to the tongue and palate. In this figure, responses to stimuli classified as sweet, salty, or sour by humans are listed on the left, whereas those classified as bitter-tasting are on the right. The neuron in these records responds robustly to stimuli of different taste qualities (e.g., NaCl and quinine). Upward arrow denotes stimulus onset.

quantitatively describe the breadth of tuning of gustatory neurons. Entropy is calculated as

$$H = -K \sum_{i=1}^{n} P_i \log P_i$$

where H is a number describing the breadth of responsiveness, K is a scaling constant, and P_i is the proportional response to each of n stimuli. The P_i for each cell is calculated as the response to the ith stimulus expressed as a proportion of the total response to all four stimuli. This metric does not depend upon a response crierion. The value of H ranges from a minimum of 0 to a maximum of 1. A neuron that responds to only one stimulus out of four would achieve $H = 0$ (i.e., no uncertainty as to which stimulus produced a response), whereas a cell that responds equally well to all stimuli would result in $H = 1$ (i.e., maximum uncertainty). This metric has seen extensive use over the past 25 years in the study of both peripheral and central gustatory neurons (see Spector and Travers[13] for a thorough review of this literature).

Using the entropy measure, it has been shown that taste neurons in the rNST are generally more broadly tuned than peripheral gustatory axons.[14] For example, chorda tympani fibers that respond maximally to sucrose relative to Na^+ salts, acidic, or bitter stimuli are the most narrowly tuned (mean $H = 0.39$, versus 0.59 for NaCl-optimal and 0.67 for HCl-optimal neurons). Yet at the level of the NST, sucrose-optimal neurons are found to be more broadly responsive ($H = 0.59$). This example shows how a single category of neuron in the periphery becomes systematically more broadly tuned at the level of the NST, likely as a result of convergent input from different kinds of peripheral fibers. Overall, cells in the central nervous system (CNS) of most species are relatively broadly responsive to stimuli representing different taste qualities.

5.2.2 CONVERGENCE OF PERIPHERAL INPUTS

Fibers of the chorda tympani nerve typically innervate taste cells in multiple fungiform papillae, although the sensitivities of each branch appear to be highly similar, suggesting that taste axons are guided to make connections to particular receptors.[15] Thus, a neuron's breadth of responsiveness does not appear to appreciably increase due to convergent input from separate axonal branches within the tongue. However, inputs from separate receptive fields are known to converge onto NST neurons,[5] resulting in greater breadth of responsiveness of these cells. What is more, rat rNST neurons have been shown to receive input from completely separate receptor populations, such as the anterior tongue and palate.[6,7]

Neurons of this sort often respond best to NaCl applied to the anterior tongue and sucrose applied to the palate. Convergence from different parts of the oral cavity onto cells in the hamster NST has also been described.[16] The increased breadth of responsiveness of NST neurons over peripheral fibers is likely due to convergent input from peripheral axons onto these cells. For example, oral application of the sodium channel blocker amiloride inhibits the robust responses to NaCl observed in sucrose-best neurons in the NST,[17,18] although some sucrose-best chorda tympani fibers are quite narrowly tuned and display poor sensitivity to NaCl relative to NaCl-best fibers, which are amiloride sensitive.

The central representation of gustatory information at the level of the rNST is arranged in an orotopic fashion as a result of the diverse anatomical distribution of taste buds across the tongue, palate, and oropharyngeal and laryngeal epithelia and the innervation of these areas by three different cranial nerves.[19,20] Afferent neurons of the VIIth nerve, including those within the chorda tympani and the greater superficial petrosal nerves, terminate in the most rostral pole of the NST. Input from taste bud fields innervated by the IXth nerve overlaps this distribution but extends more caudally. The termination of the IXth nerve is overlapped and extended even more caudally by afferent terminals of the gustatory axons of the Xth nerve. This arrangement results in the oral cavity being roughly represented spatially within the NST, with input from the anterior tongue and palate most rostral and that from the epiglottis most caudal within the nucleus.

Electrophysiological recordings taken from gustatory nerves in rodents have revealed some differences in responsiveness to different stimuli across the various receptive fields, most notably a greater sensitivity to quinine and other bitter stimuli on the posterior tongue. This difference is also reflected in the responsiveness of cells in different regions of the NST.[4,16] This regional differentiation remains at higher levels of the gustatory pathway including the gustatory cortex, where input from the VIIth and IXth nerves is still somewhat segregated.[21] Such segregation could reflect different functional roles served by these inputs. Neurotomy experiments have revealed that gustatory information provided by the VIIth nerve, but not the IXth, appears to be critical for taste discrimination,[22–24] whereas aversive taste reactivity (e.g., gapes) depends upon IXth nerve input.[25] That different nerves could serve different functional roles has been observed in other species such as fish, where the VIIth nerve is involved in food seeking and the IXth in ingestion.[26,27] Although the degree of anatomical overlap between cranial nerve inputs in mammals makes such a relationship more difficult to discern, similar functional dichotomy between VIIth and IXth nerve gustatory function may exist.

5.3 CATEGORIZING GUSTATORY NEURONS IN THE rNST

Much effort has been directed towards understanding whether there are types of gustatory neurons. Understanding this issue at the level of the rNST could provide insight into the functional organization of gustatory circuits in this structure. Different investigators have adopted different strategies to categorize gustatory neurons into purported functional groups.

5.3.1 CLASSIFICATION BY BEST-STIMULUS AND MULTIVARIATE ANALYSIS

Most studies of information processing by gustatory neurons begin by categorizing cells on the basis of their taste responsiveness. There have been multiple methods used to do this, but the most widely used classification scheme is grouping neurons by their best-stimulus, as first described by Frank[28] for taste-responsive fibers of the chorda tympani nerve. Here, chorda tympani fibers were classified into groups based on which of four stimuli representative of the basic taste qualities (sucrose, NaCl, HCl, and quinine) were most effective when presented to the anterior tongue at midrange concentrations. When the stimuli were ordered along the abscissa from most to least effective, chorda tympani fibers peaked at a single point, with profiles indicative of sucrose- (S), NaCl- (N), or HCl- (H) best fibers. Quinine-best fibers were not evident in the chorda tympani nerve, although they are commonly observed in the glossopharyngeal nerve.[29,30]

For a given neuron, if multiple stimuli are tested, one of them will undoubtedly be classified as "best." However, what is interesting is that the taste sensitivity properties of neurons within a best-stimulus group are known to be relatively similar. One way of evaluating that similarity is to use a numerical taxonomic procedure, such as hierarchical cluster analysis, to investigate the similarities within and between neural groups (e.g., Rodieck and Brening[31]). Such an analysis was applied by Frank to the responses of hamster chorda tympani fibers to 13 stimuli, resulting in three types of neurons that correspond to the S-, N-, and H-best classifications.[32]

This analysis demonstrates that neurons within a group are more similar to one another than they are to members of other groups and provides a quantitative basis for classifying neurons. Accordingly, multivariate analyses of the response profiles of hamster NST cells, even including responses from as many as 18 stimuli, resulted in classifications that were, to a large extent, predicted from the best-stimulus.[33] Earlier studies in rats, in which stimuli were not applied to the palate, did not present clear evidence for neural groups based on hierarchical clustering[34] and, in fact, argued

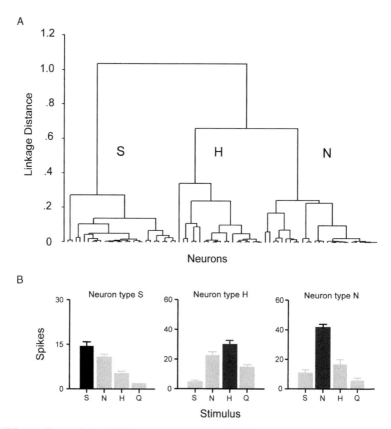

FIGURE 5.2 Categorizing NST gustatory neurons. (A) The outcome of hierarchical cluster analysis as applied to categorize 76 NST neurons into groups based on similarities among their responses to (in M) 0.5 sucrose, 0.1 NaCl, 0.01 HCl, and 0.01 quinine-HCl. In this analysis, neurons are joined into clusters based on a measure of similarity of their response profiles, defined here as the relative responses to four stimuli. Cluster analysis proceeds in a stepwise manner until all response profiles (i.e., neurons) are joined into a single cluster. The outcome of this analysis can be represented by a dendrogram (as shown in this figure), which depicts the clustering pattern and the degree of similarity (i.e., distances) among clusters. Here, the Pearson product-moment correlation was used as a measure of response profile similarity. The "within-groups linkage" amalgamation schedule was used. Cluster analysis defined three groups: S (sucrose-oriented cells), N (NaCl-oriented), and H (HCl-oriented). (B) The average responses (spikes per 1 sec of responding; bars represent standard error) of each cell group to the four stimuli. Sucrose, S; NaCl, N; HCl, H; quinine, Q.

against the existence of functional neural groups. Yet neuronal groupings derived from multivariate analysis of NST neurons in rats that correspond to S-, N-, and H-best classifications have been observed in more recent experiments where stimuli were applied to both the tongue and palate (also see Figure 5.2).[9,35,36]

5.3.2 NEURON TYPES: PHYSIOLOGICAL SIGNIFICANCE

Support for categorizing neurons into groups based on their response properties also comes from analyses of physiological parameters of gustatory fibers and neurons. For example, oral application of amiloride has been shown to inhibit responses to NaCl in mainly N-best fibers of the chorda tympani.[37] The substantial response to NaCl in HCl-best fibers is completely unaffected by amiloride treatment. This segregation of amiloride-sensitive input to the NaCl- but not the HCl-best neurons is maintained at the level of the NST.[38,39] Amiloride-sensitive N-best neurons likely play a critical role in discriminating sodium taste, as application of amiloride to the tongue disrupts the performance of rats on a two-lever operant task in which they have been taught to discriminate NaCl from KCl.[40,41] The activities of NST neurons that respond best to sweets but are also sensitive to sodium salts may also be important for the coding of sodium taste, as oral application of amiloride also inhibits NaCl responses in S-best cells.[17,18,42]

A correlate between physiology and tuning has also been observed in quinine-best neurons in the hamster NST. In a recent study, Cho et al.[43] found a substantial group of quinine-best neurons based on stimulation of the anterior tongue with equally effective concentrations of stimuli (in M): 0.032 sucrose, 0.032 NaCl, 0.0032 citric acid, and 0.032 quinine-HCl. In the hamster NST, quinine-best neurons are evident when this strong concentration of quinine is used to stimulate the anterior tongue. In this study, it was verified whether NST cells projected forward to the parabrachial nucleus, the second central synapse for taste information processing in rodents. Twenty-three quinine-best NST neurons were found that projected to the parabrachial nucleus. Interestingly, the mean conduction velocity of these quinine-best neurons was significantly lower than that of the other types of projection neurons, suggesting that the most quinine-responsive cells are a subset of smaller neurons. These data correspond well with the findings of Renehan et al.,[44] showing that neurons responding specifically to quinine in the rat NST were significantly smaller than other cells. It is worthwhile to mention that at lower concentrations of quinine, such as those used to classify chorda tympani fibers,[28] these cells would not necessarily have been classified as quinine-best, likely responding more strongly to NaCl or citric acid.

The classification of neurons by their best-stimulus does generally correspond to those categories derived from more quantitative techniques such as hierarchical cluster analysis, even when responses to more than four stimuli are analyzed. However, because of the relatively broader tuning of central taste neurons in comparison to peripheral gustatory fibers, the correspondence between a best-stimulus classification and a hierarchical cluster solution is not quite as tight in NST neurons as for peripheral fibers.[33] It is also important to consider that changes in stimulus concentration can impact the best-stimulus classification for many NST neurons.

Yet the best-stimulus method of classifying gustatory neurons has been widely used and is sometimes argued as an organizing principle in this system.

5.4 MODULATION OF ACTIVITY IN THE rNST

The response profiles of NST gustatory neurons have been shown to be dynamic rather than static attributes of these cells. That is, the sensitivities of these neurons to tastants can vary depending on a range of factors, such as the level of inhibitory synaptic input[45] to the motivational state of the organism.[46,47] Here, we discuss data regarding how physiological variables can influence the response properties of NST neurons and also neural substrates that might be involved with such modulation.

5.4.1 Physiological Modulation of rNST Activity

The initiation and control of ingestive behavior is critically mediated by gustatory input, which guides the detection and consumption of nutritive foods and promotes the avoidance of toxic substances. Thus, the responsiveness of gustatory neurons to tastants is likely a determining factor as to whether stimuli will be ingested or avoided. Various kinds of experiments have shown that physiological and experiential factors related to feeding and motivation may modulate the responsiveness of taste neurons in the NST.

5.4.1.1 Feeding and Appetite

Several studies have shown that manipulations that alter an animal's state of satiety can have an influence on the responses of gustatory neurons in the NST. For example, hyperglycemia produced by intravenous infusion of glucose decreases the sensitivity of NST gustatory neurons to orally applied glucose, with a 43% reduction in sensitivity observed on average across cells.[47] Blood glucose levels had considerably less of an effect on responses to NaCl and HCl (20% and 16% reductions, respectively) and nominal effects on responses to quinine (3%). Accordingly, intravenous infusion of glucose resulted in a decreased preference for glucose by rats but did not affect behavioral responses to quinine.[48] Thus, elevation of blood sugar levels results in decreased gustatory sensitivity to sugars in the NST, accompanied by a reduction in the intake of glucose. Intravenous infusion of other satiety factors, including insulin and pancreatic glucagon, also decreased NST multiunit responses to glucose, although cholecystokinin (CCK) had no effect on taste activity.[49] Similarly, the responsiveness of NST neurons to taste stimulation is also reduced following gastric distension.[50] Such factors could play a part in contributing to satiety and appear to decrease gustatory sensitivity to nutritive/caloric taste stimuli.

The ability to detect and consume sodium is critical for survival, and sodium appetite has been shown to modulate the responsiveness of gustatory neurons. Under normal conditions, rats will choose to ingest NaCl at isotonic concentrations. Thus, 0.17 M NaCl is avidly consumed, whereas other concentrations are less preferred or avoided, and rats can voluntarily mix NaCl and water to maintain isotonic levels of intake.[51] However, when plasma sodium concentration declines, the creation of a specific sodium appetite may serve to return sodium levels to baseline levels,[52] and taste sensitivity to salts could play an important role in this drive. Rats that are made sodium deficient will display strong preference for sodium salts, such as NaCl,[53,54] and will ingest these salts at concentrations that might otherwise be avoided. Under conditions of sodium deficiency, the sensitivity to NaCl in gustatory neurons in both the chorda tympani[55,56] and NST[57] is decreased. This reduction is particularly apparent in those neurons most responsive to sodium salts. These results have been interpreted to mean that a sodium-deprived rat would require greater ingestion of NaCl to establish the same level of sensory input as a normal rat.[56] Thus, during sodium deficiency, decreased gustatory responsiveness is associated with increased intake of NaCl.

5.4.1.2 Conditioned Avoidance

Conditioned taste aversion is an associative learning phenomenon first described by Garcia and colleagues,[58] by which a normally appetitive taste stimulus can be made aversive to a rodent by pairing this stimulus with gastrointestinal illness. Rodents will then associate the resulting illness with the taste of the conditioned stimulus (CS) and will avoid it and other stimuli with a taste similar to the CS.[59,60] This behavior has survival value for the animal by avoiding the ingestion of food that contains toxic substances. Thus, such learning would be expected to be reflected in the activities of gustatory neurons. Accordingly, Chang and Scott[61] demonstrated that aversive conditioning to saccharin increased the responsiveness to the CS in those NST neurons most responsive to sweet substances. Thus, such conditioning alters the responsiveness of gustatory neurons generally by making the CS a somewhat more effective stimulus, possibly serving the function of amplifying the neural message for the CS so that it can be reliably detected and readily avoided.

5.4.2 ELECTROPHYSIOLOGICAL ANALYSIS OF DESCENDING MODULATION

Taste information is projected from the NST to the PBN, which, in turn, sends information to two different systems: a thalamocortical pathway and limbic forebrain structures. Anatomical experiments have shown that some PBN neurons project to the ventral posterior medial nucleus parvicellularis (VPMpc) of the thalamus, whereas another subset of neurons projects to the central nucleus of the amygdala (CeA).[62] From the VPMpc, taste information is transmitted to the gustatory insular cortex. PBN neurons also project to the lateral hypothalamus (LH), bed nucleus of the stria terminalis (BST), and the substantia innominata.[63,64] Structures in the limbic forebrain gustatory pathway have been implicated in food and water regulation, sodium appetite, taste aversion learning, and the response to stress, suggesting a substrate through which taste activity interfaces with motivated behavior.

There are direct descending projections from the gustatory cortex, the lateral hypothalamus, the central nucleus of the amygdala, and the bed nucleus of the stria terminalis to the gustatory regions of the NST (Figure 5.3).[65–68] All of these forebrain

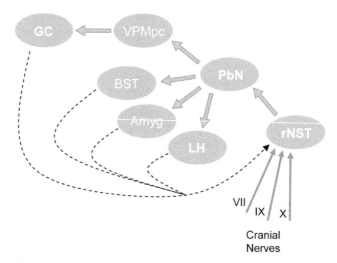

Cranial
Nerves

FIGURE 5.3 The responsiveness of NST gustatory neurons reflects not only sensory input from the taste receptors, but descending influences from forebrain circuits. Gustatory circuits in the rNST receive sensory input from cranial nerves VII, IX, and X. In rodents, taste information is then projected to the parabrachial nucleus (PBN) of the pons, which, in turn, routes gustatory information along a thalamocortical or ventral forebrain pathway. Taste information reaches the gustatory cortex (GC) via a projection from the PBN to the ventral posterior medial nucleus of the thalamus, parvocellular division (VPMpc). Structures in the ventral forebrain that receive input from the PBN include the bed nucleus of the stria terminalis (BST), central nucleus of the amygdala (Amyg), and the lateral hypothalamus (LH). The GC, BST, Amyg, and LH maintain descending connections with the brainstem and modulate taste processing in circuits in the NST.

regions of the gustatory system have been shown to produce modulation of taste activity in the NST. Thus, the responsiveness of brainstem gustatory neurons reflects not only sensory input from the taste receptors, but also descending influences from forebrain circuits. Such descending influences may constitute a neural substrate that contributes to the modulation of NST activity under altered physiological and motivational states.

5.4.2.1 Cortex

Di Lorenzo and Monroe[69] showed both increases and decreases in the responses of NST neurons to taste stimulation following blockade of activity in the gustatory cortex (GC) by local infusion of procaine, revealing that NST cells can be either facilitated or inhibited by cortical influences. Investigators employing electrical or chemical stimulation of the GC in hamsters have arrived at similar conclusions.[45] The results of a recent study showed that of 50 neurons recorded from the hamster NST, 17 (34%) were modulated by ipsilateral GC stimulation. About half of these (8/17) were inhibited, and half (9/17) were excited. Although the excitatory effects were distributed across S-, N-, and citric acid-best neurons, the inhibitory effects of cortical stimulation were significantly more common in N-best NST cells. A more recent experiment, in which stimulating electrodes were implanted bilaterally in the GC, revealed a tendency for greater modulatory influence of the contralateral cortex on NST neurons: 16 of 50 cells (32%) were found to be modulated ipsilaterally, whereas 20 of 50 (40%) responded to contralateral stimulation.[70] Further, 11 of these cells received converging modulation from both sides of the cortex. Thus, among the 50 cells recorded, 25 (50%) were modulated by one or both sides of the GC. These stimulation experiments and earlier studies in which the cortex was anesthetized with procaine demonstrate that gustatory processing in the NST is directly modulated by activity in the GC.

5.4.2.2 Lateral Hypothalamus

Neurons within the lateral hypothalamus are known to alter their activities during food ingestion[73,74] and respond to taste stimuli applied to the oral cavity.[71,72] There are descending projections from the LH to the NST,[67,68] and prior work has suggested that stimulation of the LH enhances the responsiveness of rat NST neurons to chorda tympani nerve stimulation[75] and electrical stimulation of the tongue.[76] A recent study in hamsters showed that electrical and chemical stimulation of the LH resulted in orthodromic modulation of half (49/99) of NST taste-responsive neurons.[77] The effects were predominantly excitatory, with 44 of 99 cells showing response facilitation, whereas only 6/99 cells were inhibited. Thus, LH modulation of taste-responsive cells of the NST primarily results in facilitation in responding in these cells. Given that stimulation of the LH induces feeding behavior and lesions of the LH reduce food intake,[78,79] and that these effects have been shown to interact with taste-guided behavior,[80-82] it is possible that activity in the LH could serve the function of enhancing the gain of NST taste neurons during bouts of feeding.[77] This could serve as a mechanism to increase tastant detectability.

5.4.2.3 Central Nucleus of the Amygdala

The central nucleus of the amygdala contains neurons that respond differentially to hedonically positive and negative taste stimuli,[83] and both the CeA and basolateral amygdala are involved in conditioned taste-aversion learning.[84] There is a direct descending projection from the CeA to the NST,[64,66,85] and stimulation of the CeA has been shown to alter taste processing in the rNST. A study in hamsters showed that electrical activation of the CeA orthodromically modulated 36 of 109 taste-responsive NST neurons (33%),[86] 33 of these cells were excited, and 3 were inhibited. Interestingly, those neurons that were modulated by the CeA displayed significantly smaller responses to taste stimuli than those that were not modulated by the CeA. However, there was no specific effect on any one stimulus or cell type within the NST.

Moreover, a recent investigation has shown gustatory neurons in the NST that are modulated by the CeA are often also modulated by additional forebrain structures.[87] Of 113 cells in the hamster NST that were modulated by stimulation of either the LH or the CeA, 52 of them were responsive to stimulation of both of these sites. In this study, the influence of either site was most often similar, although there were cells that were excited by one site and inhibited by another. It becomes clear from these data that there exists a complex descending influence on the gustatory responses of cells in the rNST.

5.4.2.4 Bed Nucleus of the Stria Terminalis

The dorsolateral bed nucleus of the stria terminalis sends a descending projection to the NST.[67,68,88] A recent study in hamsters revealed that stimulation of the BST resulted in inhibition of responding in 29/101 NST taste neurons, whereas only 7 were excited. All types of NST neurons were found to be affected by BST stimulation. However, the number of NaCl-best neurons was fewer and citric acid-best cells greater among those modulated by the BST than expected by chance. Although various subnuclei of the BST are implicated in a number of neural systems, including those involved in responses to stress,[89,90] and in motivation, reward, and drug addiction,[91,92] the role played by the BST in the processing of taste information is not clear. It is possible that descending input from the BST could play a role in the effects of stress on eating behavior.[93–95]

Overall, experiments described here have revealed an extensive centrifugal modulation of gustatory responsiveness in the rNST. Essentially every forebrain target of the gustatory system, including the GC, LH, CeA, and BST, exerts a certain degree of influence on the taste response properties of NST neurons. In summary, the LH and CeA exert a predominantly excitatory effect on NST taste cells, whereas the GC and particularly the BST produce significant inhibition. This extensive neural substrate no doubt underlies the modulation of taste activity by physiological and experiential factors. Future studies geared towards understanding how these pathways are engaged by alterations in blood glucose, gastric distension, conditioning, and other physiological conditions known to alter taste activity may provide further clues as to the exact function served by descending input from these structures on taste processing in the NST.

5.5 THEORIES OF TASTE QUALITY CODING

How is information about taste quality carried by the activities of gustatory neurons? The issue of gustatory quality coding has been the subject of a long-standing debate, which has largely revolved around two coding models: the labeled-line and across-neuron pattern theories. Both of these are spatial models of neural coding: Labeled-line theory postulates that information about taste quality is signalled by which cells are active, whereas across-neuron pattern theory declares that a pattern of activity across a group of cells carries information about stimulus quality. Here, we will briefly review these theories along with their basic assumptions. In Section 5.6, we will discuss evidence regarding a spatial neural code for taste, focusing on the relationship between recent developments on the molecular biology of taste receptors and the response properties of gustatory neurons in the rNST. Further, we consider how the time course of spike activity could play a role in taste coding in the rNST.

5.5.1 ACROSS-NEURON PATTERN THEORY

Early neurophysiological studies revealed that gustatory neurons in several species are sensitive to stimuli representing more than one taste quality. These data led to the idea that taste quality information is encoded by relative patterns of activity generated across a population of neurons.[96,97] This theory became known as coding by across-neuron patterns in the study of gustation. Under this model, the activity of any one cell is only meaningful when taken in the context of its neighbors.[96–99] The common conceptualization of pattern theory in gustation has been largely based on the degree of pairwise correlation between activity profiles produced by different tastants across a large number of gustatory neurons. Under this framework, discriminability between two stimuli is inversely proportional to the degree of correlation between their patterns produced across a population of cells.[100] The degree of correlation between across-neuron patterns of response does, in many cases, appear to reflect the degree of perceived similarity between taste stimuli. Stimuli that taste similar to humans and are categorized as similar by rodents in behavioral experiments evoke correlated patterns of response, whereas dissimilar stimuli produce uncorrelated across-neuron patterns.[98,99]

Across-neuron pattern theory accommodates the multiple sensitivities of taste cells and proposes that individual neurons contribute to the representation of more than one taste quality. This model assumes that a downstream processor of gustatory neurons "knows" the pattern associated with each stimulus. Only then can the pattern of activity across gustatory neurons convey meaning.[98] In its purest form, across-neuron pattern theory places no importance on the existence of neuron types. However, some studies have shown that the activities of certain types of neurons are critical for establishing different patterns of response between tastants. For example, St. John and Smith[18] used multivariate procedures to demonstrate that the activities of amiloride-sensitive N-best neurons in the rNST are critical to establish distinguishable across-neuron patterns of response to NaCl and KCl, stimuli that are behaviorally discriminable by rodents.[40]

5.5.2 LABELED-LINE THEORY

In lieu of across-neuron pattern theory, some have argued that a particular stimulus quality is encoded by the activation of one of a few discrete types of gustatory neurons. In this theory of coding, known as labeled-line, cells with a common best-stimulus are purportedly dedicated to represent the qualitative features of only this stimulus.[101,102] Labeled-line theory requires that activity within a given type of neuron is both necessary and sufficient to represent a single stimulus quality. Under this framework, a hypothetical decoder of gustatory neurons would use a binary "on/off" strategy to read out stimulus quality: When a group of neurons with a common best-stimulus is active, the decoder would report the quality of this stimulus, otherwise a stimulus of this quality is not present. Thus, sweetness, for example, is assumed to be encoded by activity in exclusively sweet-best (e.g., sucrose-best) neurons. Labeled-line theory has been largely argued to be supported by data showing that the activities of purported functional groups of neurons correlate with behavioral responding to these stimuli.[102,103]

5.6 TASTE CODING IN THE rNST

Although there has been considerable progress in recent years in our understanding of both the central neural processing and the receptor and transduction mechanisms for taste, investigators from these two perspectives view coding in the gustatory system very differently. Recent advances in molecular biology have reinvigorated the idea that taste quality may be coded by hard-wired labeled lines, which faithfully represent stimulus quality.[104–106] On the other hand, many investigators who record the activity of the broadly tuned neurons of the central nervous system favor a population code for taste quality.[3,107] As a consequence, there is still considerable controversy and debate over whether taste information is represented by specific neuron types or by the ensemble of activity across a population of broadly tuned neurons. Here, we focus on understanding the relationship between taste receptors and the response properties of neurons in the rNST.

5.6.1 Specificity of Input from Taste Receptors?

A family of genes has been discovered in mammals that accounts for much of the behavioral sensitivity towards stimuli characterized by humans as tasting sweet or savory (umami). These genes, the *Tas1rs,* encode functional G protein-coupled receptors (GPCRs) that have been demonstrated *in vitro* to bind sweet stimuli such as sugars, artificial sweeteners, and sweet proteins, or umami stimuli such as L-amino acids.[108–114] A second class of taste receptors, the T2rs, function as receptors for bitter-tasting compounds.[8,115] To date, approximately 25 human and 33 mouse functional bitter taste receptor genes (*Tas2rs*) have been identified.[116] T2rs are expressed in taste receptor cells throughout the oral cavity; individual cells tend to express multiple T2rs.[8]

Expression studies have suggested that the genes for the T1r and T2r receptors, which are involved in responses to sweet and bitter stimuli, respectively, are not co-localized within receptor cells.[117] Further, experiments in which cells normally expressing T1rs are instead made to express a T2r[118] or an opioid receptor[119] show that these taste bud cells are hardwired to signal a sweet (or palatable) taste. These authors and others[120] have suggested, therefore, that taste quality is represented by different cells in a labeled-line fashion. However, it has been known for some time that sweet and bitter stimuli are predominantly coded by different cells in the gustatory pathway.[121] Recent molecular data also suggest segregation between sweet and bitter taste in the CNS,[122] largely reflecting differences in the sensitivity of the VIIth and IXth nerves, which have a somewhat segregated central distribution.

A labeled-line code for taste would require that the purported tuning specificity observed at the level of the receptor cell be recapitulated at all levels of the nervous system. It has generally been observed that central gustatory neurons are differentially sensitive to sweet or bitter stimuli,[121] which agrees with the nonoverlapping expression patterns of receptors for these tastants. However, central sweet- or bitter-sensitive gustatory neurons including those in the rNST respond also to salts and acids,[9,123] potentially making the response of any one cell class equivocal with respect to taste quality.[3] Thus, even though different cells respond to sweet and bitter stimuli, this by itself is not definitive evidence that these cells comprise labeled lines for coding sweetness and bitterness. Moreover, an argument for a labeled-line code based on the observation that dedicated behavioral responses arise following the activation of types of taste receptor cells[124] must be cautiously evaluated, given that this phenomenon could be effectively argued to occur independently of the mechanism of gustatory coding. Although the stimulation of taste bud cells that express sweet receptors, for example, will undoubtedly result in the transmission of a "sweet" message to the brain, the perception of sweetness will follow regardless of whether this message is encoded by a labeled-line, pattern of activity, or other mechanism in the CNS (Figure 5.4). Studies of taste receptors themselves cannot address the

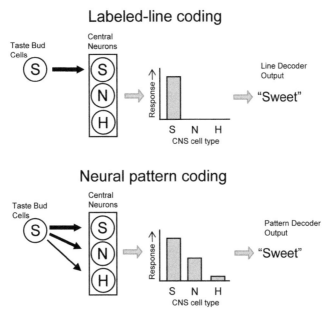

FIGURE 5.4 Dedicated behaviors that arise following stimulation of specific kinds of taste receptors do not necessarily reflect that gustatory circuits are organized according to a particular neural coding strategy. Here we hypothetically show this using the example of sweet taste. In the top panel, input from taste receptor cells that express sweet receptors ("S" cells) is transmitted to the brain in labeled-line fashion: Input from type S receptor cells is received exclusively by sweet-best (S) central neurons; NaCl- (N) or HCl- (H) best cells receive no such input. A central labeled-line decoder could then know that a sweet stimulus is present when the S line is active. In the bottom panel, input from S receptor cells is distributed across central neuron types, some of which respond best to nonsweet stimuli (N and H cells). This results in a pattern of relative activities across S, N, and H cells in the brain in response to a sweet stimulus. A central pattern decoder could recognize that a sweet stimulus is present by knowledge of this pattern. Under either coding strategy, the stimulation of sweet receptor cells results in the correct recognition of a sweet stimulus. Yet a pattern-like code would appear to be a better fit with the available data on central gustatory neurons, which are multisensitive across stimulus qualities. Under a labeled-line coding strategy, the pattern resulting from stimulation of sweet taste receptors as shown in the bottom panel could be erroneously interpreted as a stimulus that tastes sweet, salty, and sour, with the magnitude of each component proportional to the degree of activation of each channel.

organization of downstream central circuits that process input from these receptors, which more critically define the logic of gustatory neural coding.

It is worth mentioning that functional studies of taste bud cells in intact tongue and palate epithelia have shown these cells to be slightly more narrowly tuned on average than CT nerve fibers,[125] but not as specific as the molecular data on GPCRs would suggest. For example, calcium imaging and electrophysiological studies have shown that a proportion of mammalian taste bud cells responds to bitter stimuli and those of other taste qualities.[125–127] Moreover, calcium imaging studies in *Drosophila* have shown that taste receptor cells that respond to palatable sweet stimuli, such as sucrose or glucose, respond just as well to stimuli that are aversive to the fly, such as KCl or high concentrations of NaCl.[128] In this study, it was postulated that these appetitive and aversive stimuli are detected through a comparison of inputs from different kinds of taste cells. Although not argued in this paper, this hypothesis is in line with a pattern code for taste.

5.6.2 RESPONSIVENESS OF rNST NEURONS

Cells at all levels of the gustatory system, including taste bud cells,[129–131] exhibit multiple sensitivities to stimuli representing different taste qualities to humans and different behavioral categories to rodents.[32,132,133] In the periphery, sucrose-best fibers in the CT nerve[32] and QHCl-best fibers in the IXth nerve[29,30] are relatively narrowly tuned, largely reflecting the segregation of T1r and T2r receptors in the taste buds. However, due to convergence of peripheral fibers onto central neurons,[5,7,16] cells in the CNS are more broadly tuned.[14,121] Thus, whatever specificity exists at the level of the taste bud cell is degraded as distal elements converge onto more proximal ones.

In the rNST, the broad tuning of gustatory neurons in this structure questions whether individual cells or purported functional groups of them could reliably signal only a single stimulus quality.[98,99] For example, rat NST neurons that respond optimally to HCl or quinine are not differentially sensitive to these stimuli relative to Na[+] salts.[9] If neurons of this class exclusively encode an aversive/bitter taste when activated, it would be expected that NaCl would elicit a distinctly bitter taste, given that NaCl drives these acidic/bitter-sensitive neurons just as effectively as many strongly bitter stimuli.[9] Furthermore, under this type of coding strategy, NaCl would be expected to possess a sweet-taste component given that sucrose-best NST neurons receive significant input from amiloride-sensitive sodium receptors[17,18,42] that mediate the detection of sodium taste.[40] However, rodents categorize NaCl as perceptually independent of sweet, acidic, or bitter stimuli in behavioral paradigms.[103,134]

If identified neuron types in the rNST are coding stimuli in a labeled-line fashion, their broad tuning would suggest that they would do so very poorly. The broad tuning of gustatory neurons in the rNST is more conducive to a population-based model of coding,[3,98,99] where information about taste quality is carried by the activity of an ensemble of heterogeneous gustatory neurons. Information about taste quality may be contributed by the absolute pattern of response across neurons or relative levels of activation between different kinds of multisensitive cells.

5.6.3 A Temporal Component to Gustatory Coding in the rNST

Temporal coding implies that information about a stimulus is carried by the time course of spike activity. There is evidence from neurophysiological studies in both the gustatory cortex[107,135] and brainstem[136,137] that the time course of spike firing could play a role in the neural code for taste. In a recent study in the rNST, Di Lorenzo and Victor[137] used a theoretical approach to measure how spike timing, spike intervals, and spike counts in the responses of gustatory neurons could provide information about stimulus quality. Here, neurons were stimulated with individual tastants over many trials. It was found that the spike counts of some cells were highly variable from trial to trial, resulting in a changeable best stimulus in some cases. In these cells, spike count would provide equivocal information about stimulus quality. Yet spike timing was found to improve stimulus discriminability in many neurons. Interestingly, cells with the most variable spike counts tended to show the largest improvements in stimulus discriminability when temporal parameters of responses were considered.

Could both temporal and spatial parameters of neuronal responses contribute to gustatory coding? Although the timing of spikes would carry no meaning under a labeled-line strategy, it is conceivable that temporal coding could very well function in parallel with a population-based code for taste. This arrangement could provide a mechanism to increase the information-handling capacity of a group of cells, with temporal parameters providing an additional dimension to, for example, carry information that might allow for fine discriminations between stimuli. Whether the brain indeed adopts algorithms to read out gustatory information based on these parameters remains an open question.

ACKNOWLEDGMENT

The authors would like to acknowledge the support of NIH grants DC00353 and DC008194 from the National Institute of Deafness and Other Communication Disorders in the preparation of this chapter.

REFERENCES

1. Pfaffmann, C., Taste, its sensory and motivating properties, *Am. Scientist,* 52, 187, 1964.
2. Norgren, R., Taste and the autonomic nervous system, *Chem. Senses,* 10, 143, 1985.
3. Smith, D. V. and Scott, T. R., Gustatory neural coding, in *Handbook of Olfaction and Gustation,* 2nd ed., Doty, R. L. (Ed.), Marcel Dekker, New York, 2003, 731.
4. Travers, S. P., Orosensory processing in neural systems of the nucleus of the solitary tract, in *Mechanisms of Taste Transduction,* Simon, S. A. and Roper, S. D. (Eds.), CRC Press, Boca Raton, FL, 1993, 339.
5. Vogt, M. B. and Mistretta, C. M., Convergence in mammalian nucleus of solitary tract during development and functional differentiation of salt taste circuits, *J. Neurosci.,* 10, 3148, 1990.

6.   Travers, S. P., Pfaffmann, C., and Norgren, R., Convergence of lingual and palatal gustatory neural activity in the nucleus of the solitary tract, *Brain Res.*, 365, 305, 1986.
7.   Travers, S. P. and Norgren, R., Coding the sweet taste in the nucleus of the solitary tract: differential roles for anterior tongue and nasoincisor duct gustatory receptors in the rat, *J. Neurophysiol.*, 65, 1372, 1991.
8.   Adler, E. et al., A novel family of mammalian taste receptors, *Cell*, 100, 693, 2000.
9.   Lemon, C. H. and Smith, D. V., Neural representation of bitter taste in the nucleus of the solitary tract, *J. Neurophysiol.*, 94, 3719, 2005.
10. Frank, M. and Pfaffmann, C., Taste nerve fibers: a random distribution of sensitivities to four tastes, *Science*, 164, 1183, 1969.
11. Perrotto, R. S. and Scott, T. R., Gustatory neural coding in the pons, *Brain Res.*, 110, 283, 1976.
12. Smith, D. V. and Travers, J. B., A metric for the breadth of tuning of gustatory neurons, *Chem. Sens. Flav.*, 4, 215, 1979.
13. Spector, A. C. and Travers, S. P., The representation of taste quality in the mammalian nervous system, *Behav. Cognit. Neurosci. Rev.*, 4, 143, 2005.
14. Van Buskirk, R. L. and Smith, D. V., Taste sensitivity of hamster parabrachial pontine neurons, *J. Neurophysiol.*, 45, 144, 1981.
15. Oakley, B., Receptive fields of cat taste fibers, *Chem. Sens. Flav.*, 1, 431, 1975.
16. Sweazey, R. D. and Smith, D. V., Convergence onto hamster medullary taste neurons, *Brain Res.*, 408, 173, 1987.
17. Boughter, J. D., Jr. and Smith, D. V., Amiloride blocks acid responses in NaCl-best gustatory neurons of the hamster solitary nucleus, *J. Neurophysiol.*, 80, 1362, 1998.
18. St. John, S. J. and Smith, D. V., Neural representation of salts in the rat solitary nucleus: brainstem correlates of taste discrimination, *J. Neurophysiol.*, 84, 628, 2000.
19. Contreras, R. J., Beckstead, R. M., and Norgren, R., The central projections of the trigeminal, facial, glossopharyngeal and vagus nerves: an autoradiographic study in the rat, *J. Auton. Nerv. Syst.*, 6, 303, 1982.
20. Hamilton, R. B. and Norgren, R., Central projections of gustatory nerves in the rat, *J. Comp. Neurol.*, 222, 560, 1984.
21. Yamamoto, T. and Yuyama, N., On a neural mechanism for cortical processing of taste quality in the rat, *Brain Res.*, 400, 312, 1987.
22. Spector, A. C. and Grill, H. J., Salt taste discrimination after bilateral section of the chorda tympani or glossopharyngeal nerves, *Am. J. Physiol.*, 263, R169, 1992.
23. Spector, A. C., Markison, S., St. John, S. J., and Garcea, M., Sucrose vs. maltose taste discrimination by rats depends on the input of the seventh cranial nerve, *Am. J. Physiol. Regul. Integ. Comp. Physiol.*, 272, R1210, 1997.
24. St. John, S. J. and Spector, A. C., Behavioral discrimination between quinine and KCl is dependent on input from the seventh cranial nerve: implications for the functional roles of the gustatory nerves in rats, *J. Neurosci.*, 18, 4353, 1998.
25. Travers, J. B., Grill, H. J., and Norgren, R., The effects of glossopharyngeal and chorda tympani nerve cuts on the ingestion and rejection of sapid stimuli: an electromyographic analysis in the rat, *Behav. Brain Res.*, 25, 233, 1987.
26. Atema, J., Structures and functions of the sense of taste in the catfish (*Ictalurus natalis*), *Brain Behav. Evol.*, 4, 273, 1971.
27. Bardach, J. E., Todd, J. H., and Crickmer, R., Orientation by taste in fish of the genus *Ictalurus*, *Science*, 155, 1276, 1967.
28. Frank, M., An analysis of hamster afferent taste nerve response functions, *J. Gen. Physiol.*, 61, 588, 1973.

29. Hanamori, T., Miller, I. J., Jr., and Smith, D. V., Gustatory responsiveness of fibers in the hamster glossopharyngeal nerve, *J. Neurophysiol.*, 60, 478, 1988.

30. Frank, M. E., Taste-responsive neurons of the glossopharyngeal nerve of the rat, *J. Neurophysiol.*, 65, 1452, 1991.

31. Rodieck, R. W. and Brening, R. K., On classifying retinal ganglion cells by numerical methods, *Brain Behav. Evol.*, 21, 42, 1982.

32. Frank, M. E., Bieber, S. L., and Smith, D. V., The organization of taste sensibilities in hamster chorda tympani nerve fibers, *J. Gen. Physiol.*, 91, 861, 1988.

33. Smith, D. V., Van Buskirk, R. L., Travers, J. B., and Bieber, S. L., Gustatory neuron types in hamster brain stem, *J. Neurophysiol.*, 50, 522, 1983.

34. Woolston, D. C. and Erickson, R. P., Concept of neuron types in gustation in the rat, *J. Neurophysiol.*, 42, 1390, 1979.

35. Di Lorenzo, P. M., Lemon, C. H., and Reich, C. G., Dynamic coding of taste stimuli in the brainstem: effects of brief pulses of taste stimuli on subsequent taste responses, *J. Neurosci.*, 23, 8893, 2003.

36. Lemon, C. H. and Di Lorenzo, P. M., Effects of electrical stimulation of the chorda tympani nerve on taste responses in the nucleus of the solitary tract, *J. Neurophysiol.*, 88, 2477, 2002.

37. Hettinger, T. P. and Frank, M. E., Specificity of amiloride inhibition of hamster taste responses, *Brain Res.*, 513, 24, 1990.

38. Scott, T. R. and Giza, B. K., Coding channels in the taste system of the rat, *Science*, 249, 1585, 1990.

39. Giza, B. K. and Scott, T. R., The effect of amiloride on taste-evoked activity in the nucleus tractus solitarius of the rat, *Brain Res.*, 550, 247, 1991.

40. Spector, A. C., Guagliardo, N. A., and St. John, S. J., Amiloride disrupts NaCl versus KCl discrimination performance: implications for salt taste coding in rats, *J. Neurosci.*, 16, 8115, 1996.

41. Hill, D. L., Formaker, B. K., and White, K. S., Perceptual characteristics of the amiloride-suppressed sodium chloride taste response in the rat, *Behav. Neurosci.*, 104, 734, 1990.

42. Smith, D. V., Liu, H., and Vogt, M. B., Responses of gustatory cells in the nucleus of the solitary tract of the hamster after NaCl or amiloride adaptation, *J. Neurophysiol.*, 76, 47, 1996.

43. Cho, Y. K., Li, C.-S., and Smith, D. V., Gustatory projections from the nucleus of the solitary tract to the parabrachial nuclei in the hamster, *Chem. Senses*, 27, 81, 2002.

44. Renehan, W. E., Jin, Z. G., Zhang, X. G., and Schweitzer, L., Structure and function of gustatory neurons in the nucleus of the solitary tract. II. Relationships between neuronal morphology and physiology, *J. Comp. Neurol.*, 367, 205, 1996.

45. Smith, D. V. and Li, C.-S., GABA-mediated corticofugal inhibition of taste-responsive neurons in the nucleus of the solitary tract, *Brain Res.*, 858, 408, 2000.

46. Chang, F. C. and Scott, T. R., Conditioned taste aversions modify neural responses in the rat nucleus tractus solitarius, *J. Neurosci.*, 4, 1850, 1984.

47. Giza, B. K. and Scott, T. R., Blood glucose selectively affects taste-evoked activity in rat nucleus tractus solitarius, *Physiol. Behav.*, 31, 643, 1983.

48. Giza, B. K. and Scott, T. R., Blood glucose level affects perceived sweetness intensity in rats, *Physiol. Behav.*, 41, 459, 1987.

49. Giza, B. K., Scott, T. R., and Vanderweele, D. A., Administration of satiety factors and gustatory responsiveness in the nucleus tractus solitarius of the rat, *Brain Res. Bull.*, 28, 637, 1992.

50. Glenn, J. F. and Erickson, R. P., Gastric modulation of gustatory afferent activity, *Physiol. Behav.*, 16, 561, 1976.

51. Stricker, E. M., Gannon, K. S., and Smith, J. C., Thirst and salt appetite induced by hypovolemia in rats: analysis of drinking behavior, *Physiol. Behav.*, 51, 27, 1992.

52. McCaughey, S. A. and Scott, T. A., The taste of sodium, *Neurosci. Biobehav. Rev.,* 22, 663, 1998.

53. Contreras, R. J. and Hatton, G. I., Gustatory adaptation as an explanation for dietary-induced sodium appetite, *Physiol. Behav.,* 15, 569, 1975.

54. Jalowiec, J. E. and Stricker, E. M., Restoration of body fluid balance following acute sodium deficiency in rats, *J. Comp. Physiol. Psychol.,* 70, 94, 1970.

55. Contreras, R. J., Changes in gustatory nerve discharge with sodium deficiency: a single unit analysis, *Brain Res.,* 121, 373, 1977.

56. Contreras, R. and Frank, M., Sodium deprivation alters neural responses to gustatory stimuli, *J. Gen. Physiol.,* 73, 569, 1979.

57. Jacobs, K. M., Mark, G. P., and Scott, T. R., Taste responses in the nucleus tractus solitarius of sodium-deprived rats, *J. Physiol. (Lond.),* 406, 393, 1988.

58. Garcia, J. R., Kovner, R., and Green, K. F., Cue properties vs. palatability of flavors in avoidance learning, *Psychon. Sci.,* 20, 313, 1970.

59. Nachman, M., Learned aversion to the taste of lithium chloride and generalization to other salts, *J. Comp. Physiol. Psychol.,* 56, 343, 1963.

60. Tapper, D. N. and Halpern, B. P., Taste stimuli: a behavioral categorization, *Science,* 161, 708, 1968.

61. Chang, F. C. and Scott, T. R., Conditioned taste aversions modify neural responses in the rat nucleus tractus solitarius, *J. Neurosci.,* 4, 1850, 1984.

62. Voshart, K. and van der Kooy, D., The organization of the efferent projections of the parabrachial nucleus of the forebrain in the rat: a retrograde fluorescent double-labeling study, *Brain Res.,* 212, 271, 1981.

63. Norgren, R., Gustatory afferents to ventral forebrain, *Brain Res.,* 81, 285, 1974.

64. Norgren, R., Taste pathways to hypothalamus and amygdala, *J. Comp. Neurol.,* 166, 17, 1976.

65. Allen, G. V., Saper, C. B., Hurley, K. M., and Cechetto, D. F., Organization of visceral and limbic connections in the insular cortex of the rat, *J. Comp. Neurol.,* 311, 1, 1991.

66. Halsell, C. B., Differential distribution of amygdaloid input across rostral solitary nucleus subdivisions in rat, *Ann. NY Acad. Sci.,* 855, 482, 1998.

67. Moga, M. M., Herbert, H., Hurley, K. M., Yasui, Y., Gray, T. S., and Saper, C. B., Organization of cortical, basal forebrain, and hypothalamic afferents to the parabrachial nucleus in the rat, *J. Com. Neurol.,* 295, 624, 1990.

68. van der Kooy, D., Koda, L. Y., McGinty, J. F., Gerfen, C. R., and Bloom, F. E., The organization of projections from the cortex, amygdala, and hypothalamus to the nucleus of the solitary tract in rat, *J. Comp. Neurol.,* 224, 1, 1984.

69. Di Lorenzo, P. M. and Monroe, S., Corticofugal influence on taste responses in the nucleus of the solitary tract in the rat, *J. Neurophysiol.,* 74, 258, 1995.

70. Smith, D. V., Li, C. S., and Cho, Y. C., Forebrain modulation of brainstem gustatory processing, *Chem. Senses,* 30 (Suppl. 1), i176, 2004.

71. Norgren, R., Gustatory responses in the hypothalamus, *Brain Res.,* 21, 63, 1970.

72. Yamamoto, T., Matsuo, R., Kiyomitsu, Y., and Kitamura, R., Response properties of lateral hypothalamic neurons during ingestive behavior with special reference to licking of various taste solutions, *Brain Res.,* 481, 286, 1989.

73. Ono, T., Sasaki, K., Nishino, H., Fukuda, M., and Shibata, R., Feeding and diurnal-related activity of lateral hypothalamic neurons in freely behaving rats, *Brain Res.,* 373, 92, 1986.

74. Sasaki, K., Ono, T., Muramoto, K., Nishino, H., and Fukuda, M., The effects of feeding and rewarding brain stimulation on lateral hypothalamic unit activity in freely moving rats, *Brain Res.,* 322, 201, 1984.

75. Bereiter, D., Berthoud, H. R., and Jeanrenaud, B., Hypothalamic input to brain stem neurons responsive to oropharyngeal stimulation, *Exp. Brain Res.,* 39, 33, 1980.

76. Matsuo, R., Shimizu, N., and Kusano, K., Lateral hypothalamic modulation of oral sensory afferent activity in nucleus tractus solitarius neurons of rats, *J. Neurosci.,* 4, 1201, 1984.

77. Cho, Y. K., Li, C.-S., and Smith, D. V., Taste responses of neurons of the hamster solitary nucleus are enhanced by lateral hypothalamic stimulation, *J. Neurophysiol.,* 87, 1981, 2002.

78. Frank, R. A., Preshaw, R. L., Stutz, R. M., and Valenstein, E. S., Lateral hypothalamic stimulation: stimulus-bound eating and self-deprivation, *Physiol. Behav.,* 29, 17, 1982.

79. Grossman, S. P., Dacey, D., Halaris, A. E., Collier, T., and Routtenberg, A., Aphagia and adipsia after preferential destruction of nerve cell bodies in hypothalamus, *Science,* 202, 537, 1978.

80. Conover, K. L. and Shizgal, P., Competition and summation between rewarding effects of sucrose and lateral hypothalamic stimulation in the rat, *Behav. Neurosci.,* 108, 537, 1994.

81. Murzi, E., Hernandez, L., and Baptista, T., Lateral hypothalamic sites eliciting eating affect medullary taste neurons in rats, *Physiol. Behav.,* 36, 829, 1986.

82. Touzani, K. and Velley, L., Ibotenic acid lesion of the lateral hypothalamus increases preference and aversion thresholds for saccharin and alters the morphine modulation of taste, *Pharmacol. Biochem. Behav.,* 36, 585, 1990.

83. Nishijo, H., Uwano, T., Tamura, R., and Ono, T., Gustatory and multimodal neuronal responses in the amygdala during licking and discrimination of sensory stimuli in awake rats, *J. Neurophysiol.,* 79, 21, 1998.

84. Yamamoto, T., Shimura, T., Sako, N., Yasoshima, Y., and Sakai, N., Neural substrates for conditioned taste aversion in the rat, *Behav. Brain Res.,* 65, 123, 1994.

85. Veening, J. G., Swanson, L. W., and Sawchenko, P. E., The organization of projections from the central nucleus of the amygdala to brainstem sites involved in central autonomic regulation: a combined retrograde transport-immunohistochemical study, *Brain Res.,* 303, 337, 1984.

86. Li, C. S., Cho, Y. K., and Smith, D. V., Taste responses of neurons in the hamster solitary nucleus are modulated by the central nucleus of the amygdala, *J. Neurophysiol.,* 88, 2979, 2002.

87. Cho, Y. K., Li, C. S., and Smith, D. V., Descending influences from the lateral hypothalamus and amygdala converge onto medullary taste neurons, *Chem. Senses,* 28, 155, 2003.

88. Whitehead, M. C., Bergula, A., and Holliday, K., Forebrain projections to the rostral nucleus of the solitary tract in the hamster, *J. Comp. Neurol.,* 422, 429, 2000.

89. Cecchi, M., Khoshbouei, H., Javors, M., and Morilak, D. A., Modulatory effects of norepinephrine in the lateral bed nucleus of the stria terminalis on behavioral and neuroendocrine responses to acute stress, *Neuroscience,* 112, 13, 2002.

90. Herman, J. P. and Cullinan, W. E., Neurocircuitry of stress: central control of the hypothalamo-pituitary-adrenocortical axis, *Trends Neurosci.,* 20, 78, 1997.

91. Macey, D. J., Smith, H. R., Nader, M. A., and Porrino, L. J., Chronic cocaine self-administration upregulates the norepinephrine transporter and alters functional activity in the bed nucleus of the stria terminalis of the rhesus monkey, *J. Neurosci.,* 23, 12, 2003.

92. Walker, J. R., Ahmed, S. H., Gracy, K. N., and Koob, G. F., Microinjections of an opiate receptor antagonist into the bed nucleus of the stria terminalis suppress heroin self-administration in dependent rats, *Brain Res.,* 854, 85, 2000.

93. Fullerton, D. T., Getto, C. J., Swift, W. J., and Carlson, I. H., Sugar, opioids and binge eating, *Brain Res. Bull.,* 14, 673, 1985.

94. Hagan, M. M., Chandler, P. C., Wauford, P. K., Rybak, R. J., and Oswald, K. D., The role of palatable food and hunger as trigger factors in an animal model of stress-induced binge eating, *Int. J. Eat. Disord.*, 34, 183, 2003.

95. Morley, J. E. and Levine, A. S., Stress-induced eating is mediated through endogenous opiates, *Science*, 209, 1259, 1980.

96. Erickson, R. P., Sensory neural patterns and gustation, in *Olfaction and Taste*, Zotterman, Y. (Ed.), Pergamon, Oxford, U.K., 1963, 205.

97. Pfaffmann, C., The sense of taste, in *Handbook of Physiology. Sect 1. Neurophysiology*. Vol. 1, Field, J., Magoun, H. W., and Hall, V. E. (Eds.), American Physiological Society, Washington, DC, 1959, 507.

98. Scott, T. R. and Giza, B. K., Issues of gustatory neural coding: where they stand today, *Physiol. Behav.*, 69, 65, 2000.

99. Smith, D. V. and St. John, S. J., Neural coding of gustatory information, *Cur. Opin. Neurobiol.*, 9, 427, 1999.

100. Erickson, R. P., Doetsch, G. S., and Marshall, D. A., The gustatory neural response function, *J. Gen. Physiol.*, 49, 247, 1965.

101. Pfaffmann, C., Specificity of the sweet receptors of the squirrel monkey, *Chem. Senses Flav.*, 1, 61, 1974.

102. Pfaffmann, C., Frank, M., Bartoshuk, L. M., and Snell, T. C., Coding gustatory information in the squirrel monkey chorda tympani, in *Progress in Psychobiology and Physiological Psychology*, Vol. 6, Sprague, J. M. and Epstein, A. N. (Eds.), Academic Press, New York, 1976, 1.

103. Nowlis, G. H., Frank, M. E., and Pfaffmann, C., Specificity of acquired aversion to taste qualities in hamsters and rats, *J. Comp. Physiol. Psych.*, 94, 932, 1980.

104. Mueller, K. L., Hoon, M. A., Erlenbach, I., Chandrashekar, J., Zuker, C. S., and Ryba, N. J., The receptors and coding logic for bitter taste, *Nature*, 434, 225, 2005.

105. Scott, K., The sweet and the bitter of mammalian taste, *Curr. Opin. Neurobiol.*, 14, 423, 2004.

106. Zhang, Y., Hoon, M. A., Chandrashekar, J., Mueller, K. L., Cook, B., Wu, D., Zuker, C. S., and Ryba, N. J., Coding of sweet, bitter, and umami tastes: different receptor cells sharing similar signaling pathways, *Cell*, 112, 293, 2003.

107. Katz, D., Simon, S., and Nicolelis, M. A. L., Taste-specific neuronal ensembles in the gustatory cortex of awake rats, *J. Neurosci.*, 22, 1850, 2002.

108. Bachmanov, A. A., Li, X., Reed, D. R., Ohmen, J. D., Li, S., Chen, Z., Tordoff, M. G., de Jong, P. J., Wu, C., West, D. B., Chatterjee, A., Ross, D. A., and Beauchamp, G. K., Positional cloning of the mouse saccharin preference (Sac) locus, *Chem. Senses*, 26, 925, 2001.

109. Li, X., Staszewski, L., Xu, H., Durick, K., Zoller, M., and Adler, E., Human receptors for sweet and umami taste, *Proc. Natl. Acad. Sci. USA*, 99, 4692, 2002.

110. Kitagawa, M., Kusakabe, Y., Miura, H., Ninomiya, Y., and Hino, A., Molecular genetic identification of a candidate receptor gene for sweet taste, *Biochem. Biophys. Res. Commun.*, 283, 236, 2001.

111. Max, M., Shanker, Y. G., Huang, L., Rong, M., Liu, Z., Campagne, F., Weinstein, H., Damak, S., and Margolskee, R. F., Tas1r3, encoding a new candidate taste receptor, is allelic to the sweet responsiveness locus Sac, *Nat. Genet.*, 28, 58, 2001.

112. Montmayeur, J. P., Liberles, S. D., Matsunami, H., and Buck, L. B., A candidate taste receptor gene near a sweet taste locus, *Nat. Neurosci.*, 4, 492, 2001.

113. Nelson, G., Hoon, M. A., Chandrashekar, J., Zhang, Y., Ryba, N. J. P., and Zuker, C. S., Mammalian sweet taste receptors, *Cell*, 106, 381, 2001.

114. Sainz, E., Korley, J. N., Battey, J. F., and Sullivan, S. L., Identification of a novel member of the T1r family of putative taste receptors, *J. Neurochem.*, 77, 896, 2001.

115. Chandrashekar, J., Mueller, K. L., Hoon, M. A., Adler, E., Feng, L., Guo, W., Zuker, C. S., and Ryba, J. P., T2rs function as bitter taste receptors, *Cell,* 100, 703, 2000.
116. Fischer, A., Gilad, Y., Man, O., and Paabo, S., Evolution of bitter taste receptors in humans and apes, *Mol. Biol. Evol.,* 22, 432, 2005.
117. Zhang, Y., Hoon, M. A., Chandrashekar, J., Mueller, K. L., Cook, B., Wu, D., Zuker, C. S., and Ryba, N. J. P., Coding of sweet, bitter, and umami tastes: different receptor cells sharing similar signalling pathways, *Cell,* 112, 293, 2003.
118. Mueller, K. L., Hoon, M. A., Erlenbach, I., Chandrashekar, J., Zuker, C. S., and Ryba, N. J. P., The receptors and coding logic for bitter taste, *Nature,* 434, 225, 2005.
119. Zhao, G. Q., Shang, Y., Hoon, M. A., Chandrashekar, J., Erlenbach, I., Ryba, N. J. P., and Zuker, C. S., The receptors for mammalian sweet and umami taste, *Cell,* 115, 255, 2003.
120. Scott, K., The sweet and the bitter of mammalian taste, *Curr. Opin. Neurobiol.,* 14, 423, 2004.
121. Travers, J. B. and Smith, D. V., Gustatory sensitivities in neurons of the hamster nucleus tractus solitarius, *Sens. Processes,* 3, 1, 1979.
122. Sugita, M. and Shiba, Y., Genetic tracing shows segregation of taste neuronal circuitries for bitter and sweet, *Science,* 309, 781, 2005.
123. Lemon, C. H., Imoto, T., and Smith, D. V., Differential gurmarin suppression of sweet taste responses in rat solitary nucleus neurons, *J. Neurophysiol.,* 90, 911, 2003.
124. Zhao, G. Q., Zhang, Y., Hoon, M. A., Chandrashekar, J., Erlenbach, I., Ryba, N. J., and Zuker, C. S., The receptors for mammalian sweet and umami taste, *Cell,* 115, 255, 2003.
125. Gilbertson, T. A., Boughter, J. A., Jr., Zhang, H., and Smith, D. V., Distribution of gustatory sensitivities in rat taste cells: whole-cell responses to apical chemical stimulation, *J. Neurosci.,* 21, 4931, 2001.
126. Sato, T. and Beidler, L. M., Broad tuning of rat taste cells to four basic taste stimuli, *Chem. Senses,* 22, 287, 1997.
127. Caicedo, A., Kim, K. N., and Roper, S. D., Individual mouse taste cells respond to multiple chemical stimuli, *J. Physiol. (Lond.),* 544, 501, 2002.
128. Marella, S., Fischler, W., Kong, P., Asgarian, S., Rueckert, E., and Scott, K., Imaging taste responses in the fly brain reveals a functional map of taste category and behavior, *Neuron,* 49, 285, 2006.
129. Caicedo, A., Kim, K.-N., and Roper, S. D., Individual mouse taste cells respond to multiple chemical stimuli, *J. Physiol. (Lond.),* 544, 501, 2002.
130. Gilbertson, T. A., Boughter, J. D., Jr., Zhang, H., and Smith, D. V., Distribution of gustatory sensitivities in rat taste cells: whole-cell responses to apical chemical stimulation, *J. Neurosci.,* 21, 4931, 2001.
131. Ozeki, M. and Sato, M., Responses of gustatory cells in the tongue of rat to stimuli representing four taste qualities, *Comp. Biochem. Physiol. A,* 41, 391, 1972.
132. Scott, T. R. and Giza, B. K., Issues of gustatory neural coding: where they stand today, *Physiol. Behav.,* 69, 65, 2000.
133. Smith, D. V., Van Buskirk, R. L., Travers, J. B., and Bieber, S. L., Coding of taste stimuli by hamster brainstem neurons, *J. Neurophysiol.,* 50, 541, 1983.
134. Morrison, G. R., Behavioural response patterns to salt stimuli in the rat, *Can. J. Psychol.,* 21, 141, 1967.
135. Katz, D. B., Simon, S. A., and Nicolelis, M. A., Dynamic and multimodal responses of gustatory cortical neurons in awake rats, *J. Neurosci.,* 21, 4478, 2001.
136. Perrotto, R. S. and Scott, T. R., Gustatory neural coding in the pons, *Brain Res.,* 110, 283, 1976.
137. Di Lorenzo, P. M. and Victor, J. D., Taste response variability and temporal coding in the nucleus of the solitary tract of the rat, *J. Neurophysiol.,* 90, 1418, 2003.

6 Development and Plasticity of the Gustatory Portion of Nucleus of the Solitary Tract

David L. Hill and Olivia L. May

CONTENTS

6.1 INTRODUCTION

Developing mammals are highly dependent on the sense of taste to find and choose sources of food. From the earliest postnatal ages, the gustatory system plays a major role in ingestive behaviors and the proper utilization of food in conjunction with homeostatic systems. In addition to discrimination of taste quality and quantity, this sensory system must complement the needs of changing nutritive demands and age-related changes in gastrointestinal function by processing sensory information in a developmentally appropriate way. For example, age-related changes in electrolyte levels determined by gut and renal development[1] may influence age-related changes in gustatory function. Although this hypothesis has not been tested directly, changes in electrolytes, such as plasma sodium levels, occur temporally with increased peripheral taste responses during the first three postnatal weeks in rats.[2–5] Conversely, visceral function is impacted through reflexes triggered by the gustatory system. For example, preabsorptive pancreatic release of insulin in adult animals is triggered by the taste of sugar.[6] Therefore, developmental plasticity in gustatory function, especially plasticity resident to the CNS, may have wide ranging physiological effects beyond sensory coding.

Although the peripheral gustatory system is accurately portrayed as a key site of age-related and experience-related plasticity, primarily because of the ongoing turnover of adult taste bud cells,[7–9] it is clear that the nucleus of the solitary tract (NST) also shows an unexpected degree of plasticity during development. In fact, the anatomical and functional studies described in this chapter will demonstrate that the structure-function changes in this first central relay are as impressive as the cortical regions of other sensory systems. As such, the NST provides an ideal location to examine development of the local circuitry and function as well as the development of interactions with brainstem nuclei of other systems (e.g., oromotor).

It is our goal in this chapter to describe the development of the gustatory component of the NST with a particular emphasis on structural and functional plasticity. Interested readers should also examine other recent reviews with differing emphases.[10–16]

6.2 EARLY DEVELOPMENT OF AFFERENT INPUTS INTO THE NST

As described by King in Chapter 2, the gustatory NST receives afferent inputs from multiple nerves. We will focus on three of these nerve inputs: the chorda tympani nerve, the greater superficial petrosal nerve, and the glossopharyngeal nerve. These nerves innervate taste receptors on the anterior tongue (chorda tympani), palate (greater superficial petrosal), and posterior tongue (glossopharyngeal) (Figure 6.1). The soma of the chorda tympani and greater superficial petrosal nerves are contained within the geniculate ganglion, whereas soma of the glossopharyngeal nerve are in the petrosal nerve.

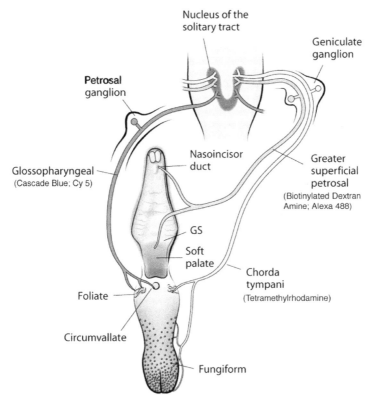

FIGURE 6.1 The anatomical organization of the peripheral gustatory system and the first central synaptic relay in the nucleus of the solitary tract (NST). The chorda tympani (CT) nerve innervates taste buds in fungiform and foliate papillae on the anterior tongue. The greater superficial petrosal (GSP) nerve innervates taste buds in the nasoincisor duct, the geschmacksstreifen (GS), and the soft palate. Both nerves comprise the VIIth cranial nerve and have cell bodies in the geniculate ganglia. The glossopharyngeal (IX) nerve innervates taste buds in foliate and circumvallate papillae on the posterior tongue, and has cell bodies in the petrosal ganglia. All three nerves terminate bilaterally in the NST; the glossopharyngeal nerve is shown on the left side only for illustrative purposes. (From May, O. L. and Hill, D. L., *J. Comp. Neurol.,* 497, 658, 2006, with permission.)

In the rat, the peak period of geniculate ganglion cell production is at embryonic day 12 (E12), whereas the peak production in the petrosal ganglion occurs at E13.[17] Programmed cell death of geniculate ganglion neurons occurs at E17;[18] no data exist on cell death of petrosal ganglia. Therefore, at least for the chorda tympani and greater superficial petrosal nerve, the adult-like number of neurons is achieved before birth (E21).

Fibers from geniculate ganglion cells arrive in the NST at about E15 and they expand significantly from E17 through E19.[19] Although these findings clearly show an early facial nerve (combined postganglionic chorda tympani and greater superficial nerves) input into the rat NST, the maturation of the chorda tympani input compared to that of the greater superficial petrosal nerve into the NST is unclear.

Moreover, the early pathfinding and guidance molecules directing gustatory nerves to their entrance into the brainstem and the NST have been unexplored. Therefore, relatively little is known about the origins of how the gustatory NST is formed.

6.3 EARLY DEVELOPMENT OF THE NST

Similar to gustatory nerves, a significant prenatal development is present for neurons resident in the NST. Presumptive target neurons for gustatory nerves in the NST are produced between E12 and E15, with the peak production occurring at E13.[17] Moreover, the cyto- and chemoarchitecture of the rat NST is well organized by birth.[20] For example, the NST becomes visible by E17 in rat, and all the subnuclei are visible by E19 (birth = E21). Acetylchoninesterase activity and a number of other biochemical markers (e.g., tyrosine hydroxylase) are mature-like by E19.

As with other central structures, it is likely that a variety of factors shape the final number of NST neurons through the coordinated regulation of cell proliferation and cell death. A prime candidate suggested to regulate the rate of cell death in the NST is insulin-like growth factor 1 (IGF-I).[21] In transgenic mice that overexpress IGF-I, there is a dramatic increase (59%) in the total volume of the medulla, which is accompanied by a 50% increase in the number of NST neurons. IGF-I has differential effects on the development of brainstem nuclei, pointing to differential expression of the molecule. This is likely one of many molecules that participate in organizing the early development of the nucleus; it is clear that significant basic cellular/molecular information is lacking in coordinating these early events. The following sections present a picture of how the NST develops after it is initially formed.

6.4 POSTNATAL DEVELOPMENT OF TERMINAL FIELDS IN THE NST

The earliest work examining the postnatal development of gustatory nerve terminal fields in the NST was done by Lasiter. Through a series of innovative experiments done over a decade ago, he found that the terminal field of the chorda tympani nerve is initially very small in the rostral NST at postnatal day 1 (P1) and then expands caudally during the next 25 days.[22] In contrast, the terminal field of the glossopharyngeal nerve does not enter its target, the intermediate zone of the NST, until P9, and expansion of the field is not complete until approximately P45.[22] The difference in terminal field development between the chorda tympani and glossopharyngeal nerves likely relates to the differential developmental time course of peripheral taste receptors.[23–25] Taste buds in fungiform papillae (located on the anterior tongue innervated by the chorda tympani nerve) develop before taste buds in circumvallate and foliate papillae (posterior tongue innervated by the glossopharyngeal nerve). Therefore, there may be an important role of stimulus-induced activity in shaping the terminal fields. Lasiter[22] did not identify the development of the greater superficial petrosal nerve terminal field; however, based on taste bud development[23–25] and the hypothesis that terminal field maturation follows receptor cell maturation, development of this terminal field should precede all others.

P15-17
Control

Adult
Control

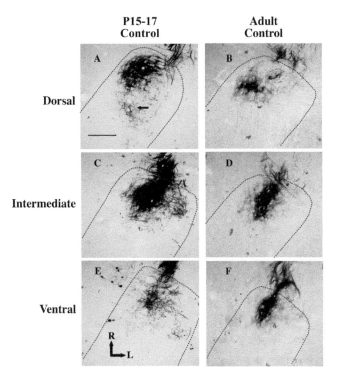

FIGURE 6.2 Photomicrographs of horizontal sections through the rostral nucleus of the solitary tract (NST) in a P15–P17-day control rat (A, C, E), and an adult control rat (B, D, F). The dorsal (A, B), intermediate (C, D), and ventral (E, F) zones of the chorda tympani nerve terminal field are shown for each rat. The chorda tympani terminal field is the dark reaction product within the rostral pole of the NST. Dashed lines outline the borders of the nucleus of the solitary tract. The scale bar in panel A denotes 200 µm and the solid arrow points to examples of axons with visible varicosities. R, rostral; L, lateral. (Reprinted from Sollars, S. I., et al., *Neuroscience*, 137, 1229, 2006, with permission from Elsevier.)

These pioneering morphological experiments have been of fundamental importance in directing further studies examining central gustatory development and organization; however, we have recent evidence strongly suggesting that many of these conclusions and developmental models must be drastically modified. In particular, unlike what was described by Lasiter and colleagues,[26,27] we found that the terminal field volume of the chorda tympani nerve is approximately 4 times greater in P15 and P25 rats compared to rats aged P35 and older.[28] That is, the field is apparently pruned from P25 to P35. As shown in Figure 6.2, the chorda tympani terminal field is large in P15 rats compared with adults, and the differences are most pronounced in the dorsal zone of the terminal field. Although the terminal field is in the proper location, sparse projections in P15 rats are seen caudal and lateral to where they are seen in adults (Figure 6.2). The composite terminal field volumes are shown in Figure 6.3. It is especially noteworthy that our data are much different from those of Lasiter and colleagues because they provide a much

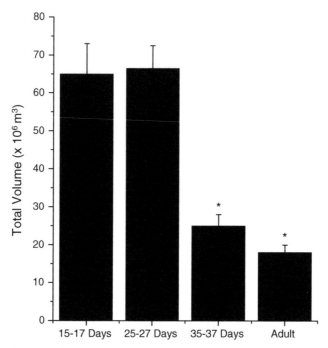

FIGURE 6.3 Total chorda tympani nerve terminal field volumes in the nucleus of the solitary tract in control rats at postnatal day 15–17, at postnatal days 25–27, at postnatal days 35–37, and at adulthood. Standard errors (SEM) are shown above the respective bars. Asterisks denote mean volumes significantly different than the preceding age group. (Reprinted from Sollars, S. I., et al., *Neuroscience*, 137, 1229, 2006, with permission from Elsevier.)

different framework from which hypotheses and interpretations are made. We show a decrease in terminal field size with age, whereas Lasiter described an increase through the first 3 weeks postnatally.[22] Therefore, we now view changes in chorda tympani nerve fields to be "curbing the excesses of youth"[29] instead of "growing" the field. Consequently, hypotheses about the underlying mechanisms acting during development (e.g., activity-dependent pruning) are much different with these data. Examples from other sensory systems demonstrate that neuronal activity shapes terminal fields[30–33] through pruning of axonal arbors.[29,34–36] This is also an attractive mechanism for altering chorda tympani terminal fields. As will be detailed in Section 6.6, there is a correspondence between the functional development of chorda tympani nerve responses and the decrease in terminal field volume. [3,4,37–39]

We have recently extended these findings by examining the terminal field development of the chorda tympani, greater superficial petrosal, and glossopharyngeal nerves in the same rat through the combined use of multiple fluorescent tracers and imaging with a confocal laser microscope.[40] This not only allows precise measurements of developing terminal fields, but it also allows an appreciation of the dynamic interrelationships among the three terminal fields with age. Quantitatively, there are major age-related changes that occur in all three terminal

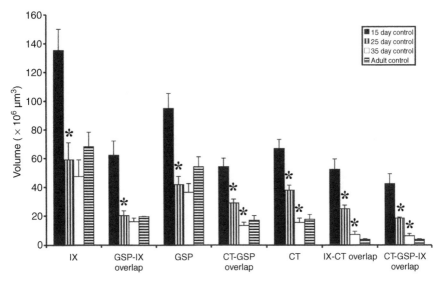

FIGURE 6.4 Mean (± SEM) total terminal field volumes of the GSP, CT, and IX nerves and corresponding overlap (e.g., GSP-CT) in the NST of control rats aged P15 (solid bars), P25 (vertically striped bars), P35 (open bars), and adults (horizontally striped bars). Total terminal field volumes decrease during normal development for all nerves and areas of overlap. Asterisks indicate mean volumes significantly lower than the earlier age ($p < 0.05$).

fields (Figure 6.4); however, there are different patterns of maturation. For the greater superficial petrosal and glossopharyngeal nerves, there is over a 50% decrease in total field volumes between P15 and P25. In contrast, chorda tympani nerve terminal field volumes decrease by 40% between P15 and P25 and then decrease again by approximately 60% between P25 and P35. It is clear, therefore, that there is widespread innervation of the NST by all gustatory nerves early in development, which is then followed by a dramatic remodeling of the field with age. Additionally, these experiments show that the potential interactions among terminal fields are also changing with age. Young rats have all fields extending more dorsal and caudal than in adults. With advancing age, there is a remodeling from where multiple field interactions are apparent, to a more restricted and mature set of terminal fields.

As noted earlier, the remodeling of chorda tympani nerve terminal fields may be related to the maturation of functional changes in afferent input. Unfortunately, support for similar mechanisms cannot be made at this time because the requisite functional developmental data do not exist for greater superficial petrosal and glossopharyngeal nerves. Although the strong possibility exists that activity plays a major role in shaping mature terminal field organization, it is possible that non-activity-dependent mechanisms also exist. For example, differences in factors such as neurotrophins[41,42] and molecular gradients such as ephrins and their receptors[43–48] may play a role in terminal field development.

6.5 MORPHOLOGICAL DEVELOPMENT OF NEURONS IN THE NST

Neurons resident in the NST also show dramatic changes in morphology. In rats, the dendritic length increases approximately threefold between P8 and P25.[27,49] There is some disagreement about dendritic branching; Lasiter et al.[27] suggest that this parameter increases with age (Figure 6.5), whereas Bao et al.[49] show that it decreases with age. In sheep, where the gustatory system develops much earlier than in the rat, dendritic lengths and dendritic spine numbers of NST neurons develop during the last trimester of gestation through early postnatal development.[50] Importantly, in both rat and sheep, maturation of NST neuron types (e.g., presumptive relay vs. interneurons) occurs along with circuit formation and neurophysiological development, further suggesting that peripheral taste responses play a role in morphological development in the NST. A major void in our understanding of NST development is the combined morphological analyses of functionally identified neurons. The only approximation to these types of analyses are experiments that examined changes in terminal field size with response characteristics of NST neurons (see Section 6.6).

Unfortunately, there are no ultrastructural studies that show developing synapses in the NST associated with identified gustatory nerves (e.g., chorda tympani inputs). However, there are limited developmental studies that examine the development of synapses, without regard to the identity of presynaptic neurons. Although synapses in the rat NST can be seen as early as E15, the earliest age at which true synapses are identified is at E19.[19] At this age, synapses contain clear, round vesicles and symmetric as well as asymmetric synapses. By birth, the density of synapses increases significantly, and synapses occur primarily on small dendrites and dendritic spines. Moreover, there is evidence of synaptic glomeruli, with clusters, which are often found in the adult NST.[19,51] Finally, multiple synaptic inputs onto single dendrites become more evident after birth.

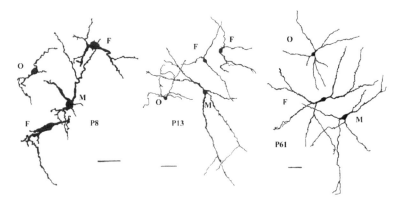

FIGURE 6.5 Representative neuron types at P8, at P13, and at adulthood (P61) from the NST viewed in horizontal sections. O, ovoid; F, fusiform; M, multipolar. Scale bars = 50 μm. (Reprinted from Lasiter, P. S., Wong D. M., and Kachele, D. L., *Brain Res. Bull.* 22, 313, 1989, with permission from Elsevier.)

Although there is an absence of developmental studies in which the identity of presynaptic elements of synapses in the NST are known, Schweitzer and colleagues have provided data on the synaptic development in which the identity of the neurotransmitter and neurotransmitter receptors are identified on NST cells. Developmentally, GABAergic cells are present in the rostral NST within the first postnatal week, but less GABA is localized in synapses than seen later in development.[52] The abundance of nonsynaptic GABA (e.g., in soma) in young animals may relate to neuritogenesis and synaptogenesis.[53] By P20, there is a shift from nonsynaptic GABA to an increase in GABAergic terminal numbers.[52] A final period of maturation occurs after weaning, during which ages the number of high-density GABAergic terminals increases.[52] A similar developmental process occurs for postsynaptic receptors for GABA.[54] Namely, $GABA_B$ receptors are diffusely distributed before P15, and they become clustered at postsynaptic sites after P15, when GABAergic synapses form. Therefore, extrasynaptic labeling for $GABA_B$ receptors decreases during development, whereas synaptic labeling increases. These studies point to circuit development in the rat NST before the second postnatal week, followed by further maturation of circuits thereafter.

6.6 FUNCTIONAL DEVELOPMENT OF NEURONS IN THE NST

Taste response development in the rat[55] and sheep[56,57] NST generally mirrors functional development of peripheral taste responses. Specifically, responses to sodium salts both in the chorda tympani nerve and in the NST increase during development. In contrast, responses to other stimuli, such as to NH_4Cl, are similar throughout development. Interestingly, fetal sheep show a lack of sodium responses in NST neurons, even though the chorda tympani nerve is responsive.[56,57] In contrast, there are small but measurable responses in P14 rats, which was the youngest age studied.[55] These examples illustrate that not all of the neural information generated in the peripheral nerves is passed onto central taste neurons during early development. Furthermore, maturation of NST responses occurs later than it does peripherally in both rat[2,3,55,58] and sheep.[56,57,59,60] In rats, mature taste responses to sodium salts in the rat chorda tympani occur at postnatal weeks 3–4, whereas mature responses appear about 2 weeks later in the NST. This implies that synaptic organization (or reorganization) directs the functional development in the NST. An excellent example of how functional changes likely reflect remodeling of primary afferent inputs comes from studies of the development of receptive fields.

Unlike chorda tympani neurons where receptive fields decrease with age,[61,62] there is an increase in receptive field sizes during development in sheep NST neurons.[63] Moreover, the receptive fields of NST neurons are larger than in chorda tympani fibers in lambs and in perinatal sheep (Figure 6.6). These results, coupled with increased spontaneous rates and increased stimulus-elicited response frequencies to NaCl, all illustrate a dynamic change in functional and structural changes at the first central gustatory relay primarily related to synaptic numbers or efficiency.[55]

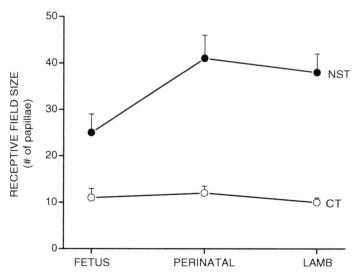

FIGURE 6.6 Receptive field sizes for NST neurons and chorda tympani fibers (CT) in fetal sheep, perinatal sheep, and lambs. Data points are mean (± SEM) field sizes. NST field sizes are larger than CT fields and NST field sizes increase significantly with age. (From Vogt, M. B. and Mistretta, C. M., *J. Neurosci.*, 10, 3148, 1990. Copyright 1990 by the Society for Neuroscience. With permission.)

The developmental delay between peripheral and central gustatory neurons cannot be attributed to poorly functionally developed NST neurons; functional membrane characteristics mature long before the taste-elicited responses. In fact, the membrane parameters of resting membrane potential, action potential, and discharge properties change the most between P5 and P15 in rats, with mature values reached by P20.[49] In addition, input resistance, membrane time constants, discharge frequency versus current pulse slopes, and afterhyperpolarization characteristics all change with age,[49] but much earlier than taste-elicited responses recorded in anesthetized rats (i.e., P50–P60). This provides further evidence that changes in the synapses and not the NST neurons per se are what account for the prolonged functional changes postnatally.

The age-related change in synaptic development is best illustrated by *in vitro* experiments examining the development of the inhibitory circuitry in the NST. *In vitro* experiments demonstrate that there are clear age-related changes that occur in the biophysical properties of IPSPs and associated pharmacological properties of NST neurons.[64] GABAergic synapses are functional at birth; however, the activity of inhibitory synapses changes significantly during the first 2 postnatal weeks. Although IPSPs in young and adult animals are hyperpolarizing, the rise time and the decay time of the IPSPs change during development.[64] In addition, there is an age-related shift in postsynaptic receptors from a combination of $GABA_A$ and $GABA_C$ receptors in young rats to $GABA_A$ receptors at adulthood. Finally, tetanic stimulation produces sustained hyperpolarization in NST neurons in rats ≤ 14 days compared to older rats.[64]

In summary, there are significant structural and functional changes that occur in the NST well beyond the period of initial innervation by the primary afferent nerves. Structurally, there are massive changes in the terminal fields. Functionally, there are dramatic changes in sodium taste sensitivities and changes in parameters consistent with increased synaptic convergence.

6.7 DEVELOPMENTAL PLASTICITY IN THE GUSTATORY NST

Research of other sensory systems demonstrates that normal maturation of function and structure depends upon proper stimulation during well-defined periods of development. In particular, the neural apparatus can be modified easily[30–33,65–74] during critical developmental periods through experimental manipulations of sensory parameters. These have been important studies not only in elucidating the capacity of the respective sensory system to respond to abnormal environmental conditions, but also in understanding the processes necessary for normal development. That is, the goal of much of this work is to learn about normal development by perturbing the system. In comparison to other sensory systems, little emphasis has been placed on clarifying the role of sensory experience in the developing gustatory system. A major focus in our laboratory has been to provide the framework for understanding how environmental events relate to gustatory development. As detailed below, we focus on the environmental effects that influence sodium taste and that induce plasticity (or lack thereof) in the central gustatory system. Other examples of how environmental effects influence NST development will follow.

To better understand the process of plasticity in the NST and to better speculate on the underlying mechanisms, we need to first present the relevant findings from the peripheral gustatory system. Restriction of maternal dietary sodium beginning on or before E8 and continued throughout development results in dramatically reduced neurophysiological responses to sodium salts in the chorda tympani nerve. Responses to NaCl are reduced by as much as 60% in sodium-restricted rats compared to controls. In contrast, taste responses to NH_4Cl and nonsalt stimuli are unaffected (Figure 6.7). The reason for the selective decrease in sodium salt-elicited responses is the absence of functional amiloride-sensitive sodium channels.[38,39,5] By using an *in vivo* voltage clamp procedure, we demonstrated that the percentage of functional sodium channels in the apical domain of taste receptor cells in restricted rats is approximately 10% of those in controls.[76] Thus, a functional taste receptor cell transduction component can be modified by early dietary manipulations.

This receptor modification, in turn, results in an altered afferent message transmitted to the central gustatory system. Importantly, the period of environmental vulnerability occurs long before the initial appearance of taste buds on the anterior tongue,[75] demonstrating that early environmental events have long-term effects on the formation of functional amiloride-sensitive sodium channels. Circulating agents such as hormones and growth factors likely play an important role in determining the response properties of developing taste receptor cells.[77] Interestingly, immunohistochemical evidence suggests that some components of the channel are present in

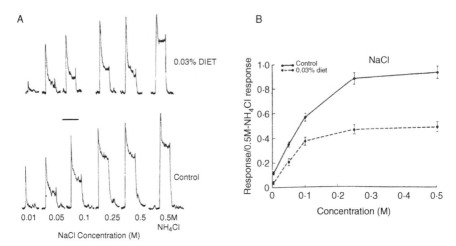

FIGURE 6.7 (A) Integrated taste responses from the chorda tympani nerve to a concentration series of NaCl (0.05 M–0.5 M) and to 0.5 M NH$_4$Cl in a sodium-restricted rat (0.03% diet) and from a control. Steady-state responses to NaCl are relatively smaller to the NH$_4$Cl response in sodium-restricted rats. Solid line denotes 20 sec. (B) Mean (± SEM) relative responses to NaCl in control (solid line) and in sodium-restricted (dashed line) rats. (From Hill, D. L., *J. Physiol. (Lond.)*, 393, 413, 1987, with permission.)

the taste receptor cells in sodium-restricted rats beginning at postnatal day 1.[78] Thus, taste receptor cells are immunohistochemically positive, even though the channel is not functional. These findings are important for understanding the plasticity in the NST, because they indicate that dietary sodium restriction throughout pre- and post-natal development produces a functional knockout-like rat for sodium taste responses. The lack of the normal functional input will then be received by the NST.

Not all populations of taste receptors are susceptible to early dietary manipulations. Palatal taste receptors, which are normally very responsive to sodium salts and to amiloride,[79] are not affected by early sodium restriction. Greater superficial petrosal nerve recordings (see Figure 6.1) reveal that early sodium-restricted rats respond normally to palatal stimulation with sodium and nonsodium salts.[80] Thus, in the same sodium-restricted rat, taste receptors on the anterior tongue have only about 10% of normal, functional amiloride channels, whereas palatal taste receptors are normal.

6.7.1 DIETARY- INDUCED ALTERATIONS IN CHORDA TYMPANI FIELDS IN THE NST

Inasmuch as the NaCl-deficient diet has such profound and selective effects on the functional development of the chorda tympani nerve, environmental alterations have even more prominent effects on the anatomical development of the NST. Dietary sodium restriction during pre- and postnatal development produces both abnormally distributed and expanded chorda tympani terminal fields.[81] In fact, an early prenatal period of sodium restriction from E3 to E12 is sufficient to produce a permanent increase in the chorda tympani terminal field.[82] (An extension of this finding will be presented in Section 6.7.8) Thus, a prominent presynaptic morphological alteration at the first central gustatory relay occurs as a result of dietary manipulations initiated during the limited time when chorda tympani and greater superficial petrosal neurons and their presumptive targets in the NST are born.[17]

The terminal field of the greater superficial petrosal nerve is not affected by early dietary sodium restriction. This field is similar to that in rats raised on a sodium-replete diet. Therefore, in the same sodium-restricted rat, the terminal field of one nerve (chorda tympani nerve) is dramatically affected, while the terminal field of another nerve that shares the same ganglion (greater superficial petrosal nerve) is unaffected. Because sodium taste responses are unaffected in the greater superficial petrosal nerve in sodium-restricted rats[83] (i.e., they are similar to control rats), the central effects mirror peripheral effects (or lack of effect). This suggests an activity-dependent mechanism that shapes the gustatory terminal fields in the NST. These topics will be presented in detail in the following sections.

6.7.2 DEVELOPMENT OF DIETARY- INDUCED ALTERATIONS IN CHORDA TYMPANI TERMINAL FIELDS IN THE NST

In all of our previous studies that examined plasticity of the chorda tympani terminal field, dietary sodium restriction was performed early in prenatal development, and the consequences of the manipulation were observed only at adulthood. It is possible that selective sodium restriction-induced expansion of the adult terminal field pattern[81] could be in place at or near the time of birth and would be present at early postnatal ages. Alternatively, the abnormal afferent terminal field may be expressed only at later postnatal ages. These two outcomes imply much different mechanisms of synaptic plasticity in the central gustatory system. To explore these potential mechanisms further, we focused on the time course of the expression of the chorda tympani terminal field development at four postnatal ages in control and sodium-restricted rats. What we found was very unexpected.

The diet-related differences in chorda tympani terminal fields occur between P25 and P35 (Figure 6.8).[28] That is, the changes reported due to early dietary manipulations[81] are not expressed morphologically until after weaning. Therefore, even though the dietary manipulation must begin early in embryonic development to have profound central morphological influences,[82] the anatomical expression of the effects does not occur until much later. What is particularly surprising was that the group-related differences are related more to the lack of terminal field reorganization during development in sodium-restricted rats and not to an abnormally expanded field, as originally thought.[83] That is, as noted earlier, there is a fourfold decrease in terminal field volume after 25 days postnatally in control rats, whereas sodium-restricted rats fail to exhibit the decrease in volume (Figure 6.8). Expressed differently, once the field expands normally by P15–P17 in sodium-restricted rats, it is "frozen" at an immature state (Figure 6.4).

To clarify the site of the decrease in chorda tympani terminal field volume, we analyzed the volumes by dividing the field into dorsal, intermediate, and ventral zones.[28] The bulk of the developmental changes occur in the dorsal zone of the

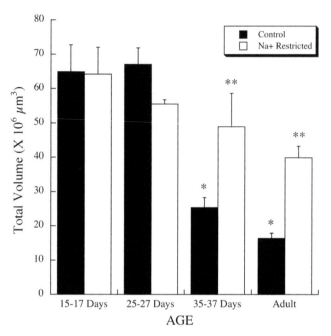

FIGURE 6.8 Total chorda tympani nerve terminal field volumes in the nucleus of the solitary tract in control (solid bars) and sodium-restricted (open bars) rats at postnatal day 15–17, postnatal days 25–27, at postnatal days 35–37, and at adulthood. Standard errors (SEM) are shown above the respective bar. Single asterisks denote mean volumes significantly different than the preceding age group of controls. Double asterisks denote significantly different means between same-aged control and sodium-restricted rats. (Reprinted from Sollars, S. I., et al., *Neuroscience*, 137, 1229, 2006, with permission from Elsevier.)

chorda tympani field (Figure 6.9). That is, the age-related and group-related differences reside almost exclusively in the dorsal portion of the chorda tympani terminal field. Due to the orientation of the NST in the brainstem, the dorsal region is also caudal. Therefore, "dorsal" refers to dorsal/caudal regions of the zone occupied by chorda tympani terminals. This is consistent with previous studies showing that the greatest amount of terminal field plasticity occurs in this zone.[22,81,82,84]

The fourfold decrease in terminal field after 25 days postnatally suggests that pruning of axonal arbors may be due to activity-dependent processes. The apparent freezing of the terminal field at an immature state in sodium-restricted rats occurs at about the time that functional sodium responses in the chorda tympani nerve are frozen.[38] Therefore, our working hypothesis is that the afferent activity responsible for the reorganization of the terminal field in controls is not present in sodium-restricted rats. The early critical period where proper dietary sodium is needed may have a direct effect on receptor cell function,[75] but may not have a direct influence (e.g., axonal pathfinding) on NST development. Similar examples from other sensory systems demonstrate that neuronal activity shapes terminal fields[85–90] through pruning of axonal arbors.[29,34–36]

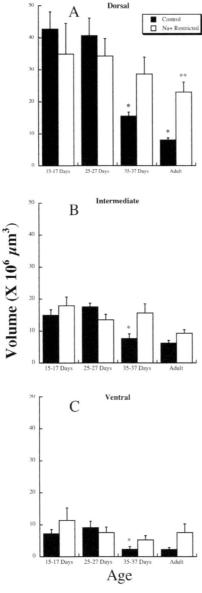

FIGURE 6.9 Chorda tympani nerve terminal field volumes for control (solid bars) and sodium-restricted (open bars) rats in the dorsal (A), intermediate (B), and ventral (C) zones of the terminal field in rats aged P15–P17 days, P25–P27 days, P35–P37 days, and adults. SEMs are shown above the respective bars. Single asterisks denote mean volumes significantly different than the preceding age group of controls. Double asterisks denote significantly different means between same-aged control and sodium-restricted rats. (Reprinted from Sollars, S. I., et al., *Neuroscience*, 137, 1229, 2006, with permission from Elsevier.)

6.7.3 The Effects of Developmental Sodium Restriction on Terminal Field Organization of Chorda Tympani, Greater Superficial Petrosal, and Glossopharyngeal Nerves in the NST

As noted earlier, we developed a triple fluorescent labeling technique used with confocal laser microscopy to fully explore the plasticity of gustatory terminal fields in the NST. This allows precise examination of the relative positions among the terminal fields of the chorda tympani, greater superficial petrosal, and glossopharyngeal nerves in the NST along with the respective volumes and precise analyses of location and volumes of overlap (see Figure 6.1 for circuitry). A clear advantage of our triple labeling approach is that, for the first time, the relative relationships among the terminal fields of three distinct gustatory nerves could be examined in the medulla of control and experimental animals. We used these techniques to uncover unexpected and dramatic relationships among three gustatory terminal fields in developmentally sodium-restricted and control rats.[91]

Developmental dietary sodium restriction results in profound alterations in the topography and distribution of two gustatory terminal fields. Early dietary sodium restriction produces a significantly enlarged chorda tympani field compared to controls (Figure 6.10). This nearly twofold enlargement occurs in the dorsal-most portion of the chorda tympani terminal field, as it expands caudally into the territories of the greater superficial petrosal and glossopharyngeal fields. The greater superficial petrosal terminal field in sodium-restricted rats is similar to controls (Figure 6.10). Finally, the total glossopharyngeal field volume in

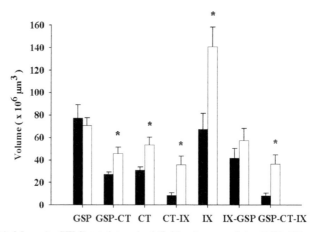

FIGURE 6.10 Mean (± SEM) total terminal field volumes of the GSP, CT, and IX nerves and corresponding overlap (e.g., GSP-CT) in the NST of control (solid bars) and sodium-restricted (open bars) rats. In restricted rats, the CT and IX fields are larger than those of controls. This expansion results in corresponding increases in the overlap among terminal fields. GSP field size does not change. Asterisks indicate $p < 0.05$. (From May, O. L., and Hill, D. L., *J. Comp. Neurol.*, 497, 658, 2006, with permission.)

restricted rats is over 2 times greater than controls (Figure 6.10). The glossopharyngeal field extends dorsally, ventrally, caudally, and laterally, well past the control glossopharyngeal field.

The amount of overlap occurring among fields in sodium-restricted rats is also different than in controls. As the chorda tympani and glossopharyngeal terminal fields in sodium-restricted rats increase in size, they expand beyond the areas normally occupied in controls. Indeed, the volume of the chorda tympani-greater superficial petrosal (CT-GSP) field overlap is significantly greater than in controls (Figure 6.10). The increase in overlapping volume is proportionate to the increase in volume of the chorda tympani field in sodium-restricted animals and does not differ between groups. The overlap between the chorda tympani and glossopharyngeal fields also significantly increases with the dietary manipulation (Figure 6.10) and is reflected in a larger proportion of the chorda tympani field, overlapping with the glossopharyngeal field. However, there are no group-related differences in the glossopharyngeal-greater superficial petrosal (IX-GSP) field overlap (Figure 6.10). Taken together, data from sodium-restricted rats indicate that the chorda tympani and glossopharyngeal fields extend past territories normally occupied in adult controls. Finally, there was a significant increase in the NST target to which all three terminal fields project (Figure 6.10), again due primarily to the increased size of the chorda tympani field.

These effects were even more dramatic when considering that the total volume of the NST in restricted rats is smaller than in controls. Furthermore, this plasticity cannot be accounted for by sheer increases in innervating afferent fibers because sodium restriction does not alter the survival of geniculate or petrosal ganglion cells.[91,92]

Figure 6.11 depicts this organization in both control and sodium-restricted rats. In control rats, the glossopharyngeal nerve terminates dorsal to the other two terminal fields and projects ventrally to overlap with the dorsal portion of the greater superficial petrosal field. The glossopharyngeal field extends more rostrally at the dorsal greater superficial petrosal zone and terminates more laterally in the NST than was previously reported.[93] The ventral portion of the glossopharyngeal field abuts the dorsal-most portion of the chorda tympani field. Chorda tympani and greater superficial petrosal fields nearly overlap completely within the rostral pole of the NST. Contrary to previous results, the coextensive chorda tympani and greater superficial petrosal fields within the rostral NST not only occupy the lateral portions of the NST, but both extend medially, as well.[92] Importantly, there is a relatively small region where all three nerves terminate. In sodium-restricted rats, the dorsal portion of the chorda tympani terminal field expands caudally outside its normal boundaries into greater superficial and glossopharyngeal field territory. The glossopharyngeal field extends dorsally, caudally, laterally, and ventrally, well past the apparent boundary of the control glossopharyngeal field. Consequently, the overlap occurring among adjacent fields also increases (Figure 6.11).

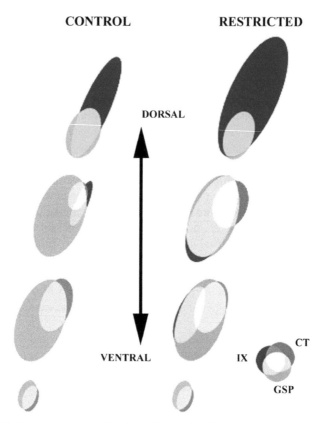

FIGURE 6.11 (See color insert following p.78) Model of terminal field organization through the dorsal-ventral extent of horizontal sections from the right NST in control and restricted rats. Overlapping fields are represented at four levels along the dorsal-ventral axis. Refer to key to identify individual fields and respective overlap among different terminal fields. Note: Due to the orientation of the NST within the brainstem, the term "dorsal" sections refers to dorsal-caudal and "ventral" refers to ventral-rostral. (From May, O. L. and Hill, D. L., *J. Comp. Neurol.*, 497, 658, 2006, with permission.)

6.7.4 THE EFFECTS OF DEVELOPMENTAL SODIUM RESTRICTION ON SYNAPTIC ORGANIZATION OF CHORDA TYMPANI, GREATER SUPERFICIAL PETROSAL, AND GLOSSOPHARYNGEAL NERVES IN THE NST

We also recently explored the synaptic alterations attendant to early dietary sodium restriction. We identified the synaptic characteristics that accompany terminal field reorganization by labeling the chorda tympani, greater superficial petrosal, or glossopharyngeal nerves in control and developmentally sodium-restricted rats, and then analyzed the ultrastructure of these nerve terminals by transmission electron microscopy.[94]

Because the dietary manipulation affects the terminal field morphology of the glossopharyngeal nerve the greatest, followed by the chorda tympani nerve,[91,92] it was predicted that the glossopharyngeal nerve would also display the most synaptic plasticity. Disruption of the greater superficial petrosal nerve terminal field morphology was not apparent at the light level; therefore, alterations in the synaptic arrangements of greater superficial petrosal nerve terminals were not expected. To the contrary, we found that dietary sodium restriction affected the synaptic connections of all three gustatory nerves. The most striking evidence for synaptic plasticity induced by dietary sodium restriction was reflected in chorda tympani axon terminal morphology (Figure 6.12). The density of chorda tympani axons and the density of synapses quadruple as a result of dietary manipulation (Figure 6.12). The most notable changes in both glossopharyngeal and greater superficial petrosal synapse morphology are an increase in synapse length and a decrease in synapse frequency compared to controls. This increased synaptic density, coupled with the decreased synaptic frequency, indicates that there are more terminals in early sodium-restricted rats compared to controls, but the synapses are spaced further apart.

Further evidence of synaptic plasticity in gustatory terminals was noted by irregularities in axon terminal morphology. Dendritic spinules, which occasionally appear in the morphological landscape of normal gustatory nerve terminals, were also often noted in sodium-restricted rats. A striking fivefold increase in these incidences occurs in chorda tympani terminals in sodium-restricted rats compared to control rats, but there is not a significant increase for greater superficial petrosal and glossopharyngeal terminals.[94]

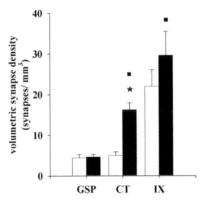

FIGURE 6.12 Volumetric density of synapses associated with GSP, CT, and IX axons in control (open bars) and sodium-restricted (closed bars) rats. Dietary sodium manipulation resulted in a significant fourfold increase in the number of synapses present on CT axons compared to controls (asterisk indicates significance). In the NST of sodium-restricted rats, significantly more synapses were associated with CT and IX axons compared to GSP axons (squares indicate significance). Error bars indicate standard error.

6.7.5 The Effects of Developmental Sodium Restriction on the Morphology of NST Cells

Prominent morphological and physiological alterations postsynaptic to gustatory nerve inputs also occur in developmentally sodium-restricted rats. Large multipolar and fusiform neurons in the rostral pole of the NST show pronounced increases in dendritic length and number, whereas there is no impact of sodium restriction on ovoid cells.[95] The neurons affected by dietary restriction are putative relay neurons.[96–98] These data strongly suggest that the morphological effects of early dietary manipulations are specific to certain cell types and may relate to their function. This conclusion is supported by work in Renehan's lab, in which changes in dendritic morphologies of NST neurons especially responsive to NaCl showed changes in dendritic orientations when dietary sodium manipulations were instituted only around and after weaning.[99] Compared to controls, sodium restriction does not affect the general appearance typical of excitatory nerve terminals (i.e., vesicle shape and asymmetry of synaptic contacts) and did not affect certain distinguishing aspects of greater superficial petrosal, chorda tympani, and glossopharyngeal nerve terminals (i.e., terminal area and postsynaptic target). However, distinct differences in the morphology and synaptic arrangements of the terminals of these three gustatory afferents are apparent, indicating a significant restructuring of neuropil.

6.7.6 The Effects of Developmental Sodium Restriction on the Functional Organization in the NST

Along with these morphological effects, the consequences of the early environmental manipulations are expressed functionally. In restricted rats, NST neurons respond with lower response frequencies to sodium salts than in controls, whereas nonsodium salts and nonsalt stimuli elicit normal responses.[100] There are also a disproportionately small number of NST neurons in sodium-restricted rats that are tuned to sodium salts.[100] Surprisingly, sodium-restricted rats fed a NaCl-replete diet for at least 5 weeks at adulthood have NST neurons that are hyper-responsive to sodium salts. This is consistent with the increased density of chorda tympani synapses in sodium-restricted rats. Recovery of peripheral function following sodium-repletion,[38,100,101] coupled with greater synaptic input, should result in greater sodium-elicited NST responses. It must be emphasized that only the anterior tongue was chemically stimulated in all of the neurophysiological studies discussed here. It is essential to do these types of experiments while stimulating distinct taste receptor populations in the mouth to build a model of central gustatory development and plasticity.

6.7.7 Other Experimental Manipulations Showing Developmental Plasticity in the NST

Although we focused much of our work on the anatomical and functional plasticity of the NST as a result of feeding a sodium-restricted diet throughout development, there are other examples that highlight how this brainstem structure changes due to early developmental manipulations.

6.7.7.1 Receptor Cell Destruction

Lasiter and colleagues used an experimental strategy similar to our sodium restriction approach by eliminating early taste experience through the destruction of taste buds on the tongue supplied by the chorda tympani nerve (i.e., fungiform papillae) by cautery. They found that destroying taste buds between P2 and P7 results in a permanent reduction in the chorda tympani terminal field by 30%.[103] The smaller terminal field is not due to fewer neurons projecting to the NST resulting from cautery, because numbers of geniculate ganglion neurons are similar to controls.[103] The authors argued that the decreased terminal field size probably reflects a receptor damage-induced change in neuronal activity during a sensitive period;[103] however, neither peripheral nor central functional responses were recorded in these rats. Therefore, it is not clear whether the early receptor cell destruction influences neuronal activity or injury-induced effects that are independent of activity. Regardless of the actual mechanism (i.e., activity vs. non-activity dependent), subsequent work indicated that the site of cellular change responsible for the central field changes resides in the geniculate ganglion and not in the terminal field.[104]

6.7.7.2 Selective Exposure to Taste Stimuli

Beginning at P4, rats were selectively presented taste stimuli for various defined periods while they were artificially reared and fed intragastrically.[26,105] The primary finding from these experiments is that limited orochemical stimulation is sufficient to produce normal axonal and terminal field development in the NST.[26,105] Exposure to water alone from P4 to P10 produces terminal fields similar to those found in rats receiving early receptor damage.[103] That is, they fail to show the caudal migration of the chorda tympani terminal field. Importantly, stimulation with either NaCl, lactose, whole rat milk, or dialyzed rat milk also produces normal fields when exposure occurs between P4 and P10[105] Finally, Lasiter[104] showed that as little as 3 days of exposure to 150 mM NaCl begun at P4 produces normal terminal fields.

These collective results are especially interesting when put in the context of the results using dietary sodium restriction throughout development. From Lasiter and colleagues' work, it appears that an absence of taste stimulation during early postnatal development results in a reduced chorda tympani terminal field, whereas early dietary sodium restriction results in an increased chorda tympani terminal field. Although these differences may be explained by when the experimental manipulation is instituted (pre- vs. postnatal), it may also be that the early development of the field is widely different. For example, we now know that early sodium restriction prevents the normal pruning of the chorda tympani terminal field.[28] It may be that early sensory stimulation of any type is necessary during early postnatal development to grow the field for subsequent pruning. Data of terminal field development in rats sustaining taste bud destruction and in rats with limited taste experience would be especially helpful in providing a basis for understanding the dynamics involved in terminal field development.

6.7.8 RECOVERY FROM EARLY DIETARY SODIUM RESTRICTION: LITTLE IS WORSE THAN MORE

We complete this section on plasticity by presenting findings that make the remarkable amount of plasticity in the gustatory NST more intriguing as well as more confusing.

Chorda tympani nerve responses recover normal function once a normal diet is restored[38] and after ingestion of one relatively large bout of physiological saline.[101,102] This functional recovery can occur at any age; therefore, the phenomenon does not have a true critical period as found in some central structures in other sensory systems.[32] These results present an interesting experimental model to examine the plasticity in the gustatory NST. Namely, do NST function and structure recover similar to peripheral taste function?

Surprisingly, sodium-restricted rats fed a NaCl-replete diet for at least 5 weeks at adulthood have NST neurons that are hyper-responsive to sodium salts. Response frequencies to sodium salts are significantly greater in recovered rats compared to controls.[100] Thus, within the 5 weeks of recovery, NST neurons change in their taste responses from being poorly responsive to sodium salts to becoming hypersensitive to these stimuli. Few functional changes occur to nonsodium stimuli.[100] These results are consistent with the increased density of chorda tympani synapses found in sodium-restricted rats.[94] The functional changes are accompanied by changes in dendritic fields of NST neurons. Multipolar and fusiform cells, which have elaborated dendritic fields in sodium-restricted rats, remain enlarged after rats are fed the NaCl-replete diet for 2–3 months.[95] However, ovoid cells show remarkable increases in the number and length of their dendrites only upon restoration of dietary sodium.[95] Multipolar and fusiform cells are putative relay neurons, whereas ovoid cells are viewed as interneurons.[96,97,106] These data strongly suggest that the morphological effects of early dietary manipulations are specific to certain cell types and may relate to their function. Although these data are striking because of the degree and specificity of functional and structural changes, we believe the most interesting (and puzzling) results from recovered rats derive from terminal field experiments.

Restoration of a sodium-replete diet in rats sodium restricted from E3 to E28 for at least 60 days results in chorda tympani nerve terminal fields that are 3 times the volume of those found in controls.[81] The mechanism attributed to this phenomenon was related to growth of the rat and of the brain when repleted.[81] That is, it is hypothesized that the relatively fixed chorda tympani terminal field is stretched along the longest axis of growth (anterior–posterior) as the rat grows. We later found that shorter periods of dietary sodium followed by sodium repletion results in even larger chorda tympani terminal fields and widespread changes in greater superficial petrosal and glossopharyngeal nerve terminal fields.

A prenatal period of sodium restriction from E3 to E12 is sufficient to produce a permanent alteration in the chorda tympani terminal field.[82] In fact, this early period of sodium restriction results in a larger chorda tympani field compared to rats fed the restricted diet from E3 to P28 before being fed the sodium-replete diet.[107] Surprisingly, the E3 to E12 period of sodium restriction also results in an approximate threefold increase in the glossopharyngeal nerve terminal field and an approximate twofold increase in the greater superficial nerve terminal field compared to controls.[107] The area of overlap relative to the individual nerve terminal fields also increases. Therefore, a 9-day period of early sodium restriction at about the time that the NST and gustatory nerve ganglia are formed results in a widespread reorganization of the presynaptic inputs into the gustatory NST. These data challenge the hypothesis that neural activity plays the primary role in reorganization of terminal fields because the sodium-replete diet was instituted even before peripheral and central gustatory structures were established.

6.8 SUMMARY AND FUTURE DIRECTIONS

The development of the gustatory NST is remarkable in that multiple sensory nerves must find their way to the appropriate level of the nucleus and expand their terminal fields in a manner that is presumably matched to the functional and behavioral repertoire of the animal, and then they must collectively remodel as the animal matures. With this dynamic system, the individual terminal fields and associated afferent messages must interact with each other to reliably relay messages about stimulus quality and quantity to the postsynaptic cell in the NST. Such a relay of information appears to involve specific rules in remodeling of synaptic inputs. All of this is done in the context of altered morphologies of postsynaptic cells with age, which could be correlational or causative with corresponding changes in the gustatory inputs. Somewhat surprisingly, the coordinated structural and functional development of the gustatory NST can be derailed by environmental events that occur even in early gestation. Unlike other sensory systems in which large changes occur only in cortical regions, the gustatory system shows a high degree of change on the presynaptic side of the first central synapse (i.e., in the NST).

We believe that the dynamics described here relating to normal development and plasticity of the NST provide a background for many future experiments. The possibilities include identifying the molecules and the timing of molecular expression used in nerve/target interactions in the nucleus, characterizing the development of circuit formation with intrinsic cells and descending projections, comprehensively examining the functional response interactions among multiple gustatory nerve inputs into the NST neurons with age, and determining how such interactions occur in awake, behaving animals. Therefore, there is an extensive array of molecular, cellular, and biophysical studies that could be done to follow up on previous developmental studies. It is certain that a better understanding of the developmental processes involved in gustatory NST function and structure will provide important insights into how the mature NST functions as an important gustatory relay and as a crucial structure that interacts with other homeostatic systems.

ACKNOWLEDGMENT

The preparation of this chapter was supported in part by NIH grants DC00407, DC06332, and DC006938 from the National Institute on Deafness and Other Communication Disorders.

REFERENCES

1. Spitzer, A., *The Kidney during Development: Morphology and Function*, Masson, New York, 1982.
2. Ferrell, M. F., Mistretta, C. M., and Bradley, R. M., Development of chorda tympani taste responses in rat, *J. Comp. Neurol.*, 198, 37, 1981.
3. Hill, D. L. and Almli, C. R., Ontogeny of chorda tympani nerve responses to gustatory stimuli in the rat, *Brain Res.*, 197, 27, 1980.
4. Yamada, T., Chorda tympani responses to gustatory stimuli in developing rats, *Jpn. J. Physiol.*, 30, 631, 1980.
5. Jelinek, J., The development of the regulation of water metabolism. V. Changes in the content of water, potassium, sodium, and chloride in the body and body fluids of rats during development, *Physiologica Bohemoslovia*, 10, 249, 1961.
6. Berridge, K., Grill, H. J., and Norgren, R., Relation of consummatory responses and preabsorptive insulin release to palatability and learned taste aversions, *J. Comp. Physiol. Psychol.*, 95, 363, 1981.
7. Beidler, L. M. and Smallman, R. L., Renewal of cells within taste buds, *J. Cell Biol.*, 27, 263, 1965.
8. Farbman, A. I., Renewal of taste bud cells in rat circumvallate papillae, *Cell Tissue Kinetics*, 13, 349, 1980.
9. Hendricks, S. J., Brunjes, P. C., and Hill, D. L., Taste bud cell dynamics during normal and sodium-restricted development, *J. Comp. Neurol.*, 472, 173, 2004.
10. Hill, D. L. and Mistretta, C. M., Developmental neurobiology of salt taste sensation, *Trends Neurosci.*, 13, 188, 1990.
11. Stewart, R. E., DeSimone, J. A., and Hill, D. L., New perspectives in a gustatory physiology: transduction, development, and plasticity, *Am. J. Physiol.*, 272, C1, 1997.
12. Mistretta, C. M., Developmental neurobiology of the taste system, in *Smell and Taste in Health and Disease*, Getchell, T. V., Doty, R. L., Bartoshuk, L. M., and Snow, J. B. (Eds.), Raven Press, New York, 1991, 35.
13. Mistretta, C. M. and Hill, D. L., Development of the taste system: basic neurobiology, in *Handbook of Clinical Olfaction and Gustation*, Doty, R. L. (Ed.), Marcel Dekker, New York, 1995, 635.
14. Stewart, R. E. and Hill, D. L., The developing gustatory system: functional, morphological and behavioral perspectives, in *Mechanisms of Taste Perception*, Simon, S. A. and Roper, S. (Eds.), CRC Press, Boca Raton, FL, 1993, 127.
15. Hill, D. L., Taste development, in *Handbook of Behavioral Neurobiology: Developmental Psychobiology*, Blass, E. M. (Ed.), Kluwer/Plenum, New York, 2001, 517.
16. Mistretta, C. M. and Hill, D. L., Development of the taste system: basic neurobiology, in *Handbook of Olfaction and Gustation*, Doty, R. L. (Ed.), Marcel Dekker, New York, 2003, 759.
17. Altman, J. and Bayer, S., Development of the cranial nerve ganglia and related nuclei in the rat, *Adv. Anat. Embryol. Cell Biol.*, 74, 1, 1982.
18. Carr, V. M., Sollars, S. I., and Farbman, A., Neuronal death in the developing rat geniculate ganglion, *Chem. Senses*, 30, A162, 2005.
19. Zhang, L. L. and Ashwell, K. W., The development of cranial nerve and visceral afferents to the nucleus of the solitary tract in the rat, *Anat. Embryol. (Berl.)*, 204, 135, 2001.
20. Zhang, L. L. and Ashwell, K. W., Development of the cyto- and chemoarchitectural organization of the rat nucleus of the solitary tract, *Anat. Embryol. (Berl.)*, 203, 265, 2001.
21. Dentremont, K. D., Ye, P., D'Ercole, A. J., and O'Kusky, J. R., Increased insulin-like growth factor-I (IGF-I) expression during early postnatal development differentially

increases neuron number and growth in medullary nuclei of the mouse, *Brain Res. Dev. Brain Res.,* 114, 135, 1999.

22. Lasiter, P. S., Postnatal development of gustatory recipient zones within the nucleus of the solitary tract, *Brain Res. Bull.,* 28, 667, 1992.

23. Mistretta, C. M., Topographical and histological study of the developing rat tongue, palate and taste buds, in *Third Symposium on Oral Sensation and Perception,* Bosma, J. F. (Ed.), Thomas, Springfield, IL, 1972, 163.

24. Hosley, M. A. and Oakley, B., Postnatal development of the vallate papilla and taste buds in rats, *Anat. Rec.,* 218, 216, 1987.

25. Harada, S. et al., Maturation of taste buds on the soft palate of the postnatal rat, *Physiol. Behav.,* 68, 333, 2000.

26. Lasiter, P. S., Effects of orochemical stimulation on postnatal development of gustatory recipient zones within the nucleus of the solitary tract, *Brain Res. Bull.,* 38, 1, 1995.

27. Lasiter, P. S., Wong, D. M., and Kachele, D. L. Postnatal development of the rostral solitary nucleus in rat: dendritic morphology and mitochondrial enzyme activity, *Brain Res. Bull.,* 22, 313, 1989.

28. Sollars, S. I. et al., Age-related decrease of the chorda tympani nerve terminal field in the nucleus of the solitary tract is prevented by dietary sodium restriction during development, *Neuroscience,* 137, 1229, 2006.

29. Kantor, D. B. and Kolodkin, A. L., Curbing the excesses of youth: molecular insights into axonal pruning, *Neuron,* 38, 849, 2003.

30. Cummings, D. M. and Brunjes, P. C., The effects of variable periods of functional deprivation on olfactory bulb development in rats, *Exp. Neurol.,* 148, 360, 1997.

31. Deitch, J. S. and Rubel, E. W., Afferent influences on brain stem auditory nuclei of the chicken: time course and specificity of dendritic atrophy following deafferentation, *J. Comp. Neurol.,* 229, 66, 1984.

32. Hubel, D. H. and Wiesel, T. N., The period of susceptibility to the physiological effects of unilateral eye closure in kittens, *J. Physiol. (Lond.),* 206, 419, 1970.

33. Renehan, W. E., Rhoades, R. W., and Jacquin, M. F., Structure-function relationships in rat brainstem subnucleus interpolaris: VII. Primary afferent central terminal arbors in adults subjected to infraorbital nerve section at birth, *J. Comp. Neurol.,* 289, 493, 1989.

34. Bagri, A. et al., Stereotyped pruning of long hippocampal axon branches triggered by retraction inducers of the semaphorin family, *Cell,* 113, 285, 2003.

35. Lichtman, J. W. and Colman, H., Synapse elimination and indelible memory, *Neuron,* 25, 269, 2000.

36. Watts, R. J., Hoopfer, E. D., and Luo, L., Axon pruning during *Drosophila* metamorphosis: evidence for local degeneration and requirement of the ubiquitin-proteasome system, *Neuron,* 38, 871, 2003.

37. Ferrell, M. F., Mistretta, C. M., and Bradley, R. M., Development of chorda tympani taste responses in rat, *J. Comp. Neurol.,* 198, 37, 1981.

38. Hill, D. L., Susceptibility of the developing rat gustatory system to the physiological effects of dietary sodium deprivation, *J. Physiol. (Lond.),* 393, 413, 1987.

39. Hill, D. L., Mistretta, C. M., and Bradley, R. M., Effects of dietary NaCl deprivation during early development on behavioral and neurophysiological taste responses, *Behav. Neurosci.,* 100, 390, 1986.

40. Mangold, J. E. and Hill, D. L., Postnatal development of gustatory nerve terminal fields in control rats, *Chem. Senses,* 30, A126, 2005.

41. Snider, W. D., Functions of the neurotrophins during nervous system development: what the knockouts are teaching us, *Cell,* 77, 627, 1994.

42. Huang, E. J. and Reichardt, L. F., Neurotrophins: roles in neuronal development and function, *Annu. Rev. Neurosci.,* 24, 677, 2001.

43. Person, A. L., Cerretti, D. P., Pasquale, E. B., Rubel, E. W., and Cramer, K. S., Tonotopic gradients of Eph family proteins in the chick nucleus laminaris during synaptogenesis, *J. Neurobiol.,* 60, 28, 2004.

44. Hansen, M. J., Dallal, G. E., and Flanagan, J. G., Retinal axon response to ephrin-As shows a graded, concentration-dependent transition from growth promotion to inhibition, *Neuron,* 42, 717, 2004.

45. King, C., Lacey, R., Rodger, J., Bartlett, C., Dunlop, S., and Beazley, L., Characterisation of tectal ephrin-A2 expression during optic nerve regeneration in goldfish: implications for restoration of topography, *Exp. Neurol.,* 187, 380, 2004.

46. Yates, P. A. et al., Topographic-specific axon branching controlled by ephrin-As is the critical event in retinotectal map development, *J. Neurosci.,* 21, 8548, 2001.

47. Prakash, N. et al., Malformation of the functional organization of somatosensory cortex in adult ephrin-A5 knock-out mice revealed by *in vivo* functional imaging, *J. Neurosci.,* 20, 5841, 2000.

48. Goodhill, G. J. and Richards, L. J., Retinotectal maps: molecules, models and misplaced data, *Trends Neurosci.,* 22, 529, 1999.

49. Bao, H., Bradley, R. M., and Mistretta, C. M., Development of intrinsic electrophysio-logical properties in neurons from the gustatory region of rat nucleus of solitary tract, *Brain Res. Dev. Brain Res.,* 86, 143, 1995.

50. Mistretta, C. M. and Labyak, S. E., Maturation of neuron types in nucleus of solitary tract associated with functional convergence during development of taste circuits, *J. Comp. Neurol.,* 345, 359, 1994.

51. Whitehead, M. C. and Frank, M. E., Anatomy of the gustatory system in the hamster: central projections of the chorda tympani and the lingual nerve, *J. Comp. Neurol.,* 220, 378, 1983.

52. Brown, M., Renehan, W. E., and Schweitzer, L., Changes in GABA-immunoreactivity during development of the rostral subdivision of the nucleus of the solitary tract, *Neuroscience,* 100, 849, 2000.

53. Meier, E., Drejer, J., and Schousboe, A., Trophic actions of GABA on the development of physiologically active GABA receptors, *Adv. Biochem. Psychopharmacol.,* 37, 47, 1983.

54. Heck, W. L., Renehan, W. E., and Schweitzer, L., Redistribution and increased specificity of GABA(B) receptors during development of the rostral nucleus of the solitary tract, *Int. J. Dev. Neurosci.,* 19, 503, 2001.

55. Hill, D. L., Bradley, R. M., and Mistretta, C. M., Development of taste responses in rat nucleus of solitary tract, *J. Neurophysiol.,* 50, 879, 1983.

56. Bradley, R. M. and Mistretta, C. M., Developmental changes in neurophysiological taste responses from the medulla in sheep, *Brain Res.,* 191, 21, 1980.

57. Mistretta, C. M. and Bradley, R. M., Taste responses in sheep medulla: changes during development, *Science,* 202, 535, 1978.

58. Yamada, T., Chorda tympani responses to gustatory stimuli in developing rats, *Jpn. J. Physiol.,* 30, 631, 1980.

59. Bradley, R. M. and Mistretta, C. M., The gustatory sense in fetal sheep during the last third of gestation, *J. Physiol.,* 231, 271, 1973.

60. Mistretta, C. M. and Bradley, R. M., Neural basis of developing salt taste sensation: response changes in fetal, postnatal and adult sheep, *J. Comp. Neurol.,* 215, 199, 1983.

61. Nagai, T., Mistretta, C. M., and Bradley, R. M., Developmental decrease in the size of peripheral receptive fields of single chorda tympani fibers and relation to increasing NaCl taste sensitivity, *J. Neurosci.,* 8, 64, 1988.

62. Mistretta, C. M., Development of receptive fields of chorda tympani nerve fibers and relation to salt taste sensation, in *The Beidler Symposium on Taste and Smell: A Festschrift*, Miller, I. J., Jr. (Ed.), Book Service Associates, Winston-Salem, NC, 1988, 21.

63. Vogt, M. B. and Mistretta, C. M., Convergence in mammalian nucleus of solitary tract during development and functional differentiation of salt taste circuits, *J. Neurosci.,* 10, 3148, 1990.

64. Grabauskas, G. and Bradley, R. M., Postnatal development of inhibitory synaptic transmission in the rostral nucleus of the solitary tract, *J. Neurophysiol.,* 85, 2203, 2001.

65. Renehan, W. E. et al., Physiological and anatomical consequences of infraorbital nerve transection in the trigeminal ganglion and trigeminal spinal tract of the adult rat, *J. Neurosci.,* 9, 548, 1989.

66. Renehan, W. E., Crissman, R. S., and Jacquin, M. F., Primary afferent plasticity following partial denervation of the trigeminal brainstem nuclear complex in the postnatal rat, *J. Neurosci.,* 14, 721, 1994.

67. Henderson, T. A., Woolsey, T. A., and Jacquin, M. F., Infraorbital nerve blockade from birth does not disrupt central trigeminal pattern formation in the rat, *Brain Res. Dev. Brain Res.,* 66, 146, 1992.

68. Buonomano, D. V. and Merzenich, M. M., Cortical plasticity: from synapses to maps, *Annu. Rev. Neurosci.,* 21, 149, 1998.

69. Catalano, S. M. and Shatz, C. J., Activity-dependent cortical target selection by thalamic axons, *Science,* 281, 559, 1998.

70. Rauschecker, J. P., Making brain circuits listen, *Science,* 285, 1686, 1999.

71. Fox, K., Wallace, H., and Glazewski, S., Is there a thalamic component to experience-dependent cortical plasticity? *Philos. Trans. R. Soc. Lond. B Biol. Sci.,* 357, 1709, 2002.

72. Katz, L. C. and Crowley, J. C., Development of cortical circuits: lessons from ocular dominance columns, *Nat. Rev. Neurosci.,* 3, 34, 2002.

73. Yan, J., Canadian Association of Neuroscience Review: development and plasticity of the auditory cortex, *Can. J. Neurol. Sci.,* 30, 189, 2003.

74. Mistretta, C. M. and Bradley, R. M., Effects of early sensory experience on brain and behavioral development, in *Studies on the Development of Behavior and the Nervous System*, Gottlieb, G. (Ed.), Academic Press, New York, 1978, 215.

75. Hill, D. L. and Przekop, P. R., Jr., Influences of dietary sodium on functional taste receptor development: a sensitive period, *Science,* 241, 1826, 1988.

76. Ye, Q., et al., Dietary Na(+)-restriction prevents development of functional Na+ channels in taste cell apical membranes: proof by *in vivo* membrane voltage perturbation, *J. Neurophysiol.,* 70, 1713, 1993.

77. Stewart, R. E., Parsons, R. J., and Hill, D. L., Development of some early sensorimotor behaviors in sodium-restricted rats, *Physiol. Behav.,* 53, 813, 1993.

78. Stewart, R. E., Lasiter, P. S., Benos, D. J., and Hill, D. L., Immunohistochemical correlates of peripheral gustatory sensitivity to sodium and amiloride, *Acta. Anat.,* 153, 310, 1995.

79. Sollars, S. I. and Hill, D. L., Taste responses in the greater superficial petrosal nerve: substantial sodium salt and amiloride sensitivities demonstrated in two rat strains, *Behav. Neurosci.,* 112, 991, 1998.

80. Sollars, S. I. and Hill, D. L., Lack of functional and morphological susceptibility of the greater superficial petrosal nerve to developmental dietary sodium restriction, *Chem. Senses,* 25, 719, 2000.

81. King, C. T. and Hill, D. L., Dietary sodium chloride deprivation throughout development selectively influences the terminal field organization of gustatory afferent fibers projecting to the rat nucleus of the solitary tract, *J. Comp. Neurol.,* 303, 159, 1991.

82. Krimm, R. F. and Hill, D. L., Early prenatal critical period for chorda tympani nerve terminal field development, *J. Comp. Neurol.,* 378, 254, 1997.

83. Sollars, S. I. and Hill, D. L., Lack of functional and morphological susceptibility of the greater superficial petrosal nerve to developmental dietary sodium restriction, *Chem. Senses,* 25, 719, 2000.

84. Pittman, D. W. and Contreras, R. J., Dietary NaCl influences the organization of chorda tympani neurons projecting to the nucleus of the solitary tract in rats, *Chem. Senses,* 27, 333, 2002.

85. Cline, H. T. (1998). Topographic maps: developing roles of synaptic plasticity, *Curr. Biol.,* 8, R836, 1998.

86. Tao, H., Zhang, L. I., Bi, G., and Poo, M., Selective presynaptic propagation of long-term potentiation in defined neural networks, *J. Neurosci.,* 20, 3233, 2000.

87. Tao, H. W., Zhang, L. I., Engert, F., and Poo, M., Emergence of input specificity of LTP during development of retinotectal connections *in vivo, Neuron,* 31, 569, 2001.

88. Zhang, L. I. and Poo, M. M., Electrical activity and development of neural circuits, *Nat. Neurosci.,* 4 Suppl., 1207, 2001.

89. Zhang, L. I., Tao, H. W., and Poo, M., Visual input induces long-term potentiation of developing retinotectal synapses, *Nat. Neurosci.,* 3, 708, 2000.

90. Zhang, L. I., et al., A critical window for cooperation and competition among developing retinotectal synapses, *Nature,* 395, 37, 1998.

91. May, O. L. and Hill, D. L., Gustatory terminal field organization and developmental plasticity in the nucleus of the solitary tract revealed through triple fluorescent labeling, *J. Comp. Neurol.,* 497, 658, 2006.

92. May, O. L. and Hill, D. L., Organization and plasticity of gustatory nerve terminal fields revealed by triple fluorescent labeling, *Chem. Senses,* 28, A42, 2003.

93. Hamilton, R. B. and Norgren, R., Central projections of gustatory nerves in the rat, *J. Comp. Neurol.,* 222, 560, 1984.

94. May, O. L., Erisir, A., and Hill, D. L., A comparative analysis of the ultrastructural morphology of the three gustatory nerve axons in the nucleus of the solitary tract in developmentally sodium-restricted and control rats, *Chem. Senses,* 30, A167, 2005.

95. King, C. T. and Hill, D. L., Neuroanatomical alterations in the rat nucleus of the solitary tract following early maternal NaCl deprivation and subsequent NaCl repletion, *J. Comp. Neurol.,* 333, 531, 1993.

96. Lasiter, P. S. and Kachele, D. L., Organization of GABA and GABA-transaminase containing neurons in the gustatory zone of the nucleus of the solitary tract, *Brain Res. Bull.,* 21, 623, 1988.

97. Whitehead, M. C., Anatomy of the gustatory system in the hamster: synaptology of facial afferent terminals in the solitary nucleus, *J. Comp. Neurol.,* 244, 72, 1986.

98. Davis, B. J. and Jang, T., Tyrosine hydroxylase-like and dopamine beta-hydroxylase-like immunoreactivity in the gustatory zone of the nucleus of the solitary tract in the hamster: light- and electron-microscopic studies, *Neurosci.,* 27, 949, 1988.

99. Liu, Y.-Z., Schweitzer, L., and Renehan, W., The influence of a modified salt diet on dendritic remodeling in the rostral nucleus of the solitary tract (rNST) of the rat, *Chem. Senses,* 24, 548, 1999.

100. Vogt, M. B. and Hill, D. L., Enduring alterations in neurophysiological taste responses after early dietary sodium deprivation, *J. Neurophysiol.,* 69, 832, 1993.

101. Przekop, P., Jr., Mook, D. G., and Hill, D. L., Functional recovery of the gustatory system after sodium deprivation during development: how much sodium and where? *Am. J. Physiol.,* 259, R786, 1990.

102. Stewart, R. E. and Hill, D. L., Time course of saline-induced recovery of the gustatory system in sodium-restricted rats, *Am. J. Physiol.,* 270, R704, 1996.
103. Lasiter, P. S. and Kachele, D. L., Effects of early postnatal receptor damage on development of gustatory recipient zones within the nucleus of the solitary tract, *Brain Res. Dev. Brain Res.,* 55, 57, 1990.
104. Lasiter, P. S. and Bulcourf, B. B., Alterations in geniculate ganglion proteins following fungiform receptor damage, *Brain Res. Dev. Brain Res.,* 89, 289, 1995.
105. Lasiter, P. S. and Diaz, J., Artificial rearing alters development of the nucleus of the solitary tract, *Brain Res. Bull.,* 29, 407, 1992.
106. Davis, B. J. and Jang, T., A Golgi analysis of the gustatory zone of the nucleus of the solitary tract in the adult hamster, *J. Comp. Neurol.,* 278, 388, 1988.
107. Mangold, J. E and Hill, D. L., Differential postnatal development of gustatory nerve terminal fields in control rats and E3-E12 sodium-restricted rats, *Chem. Senses,* 31, in press, 2006.

7 rNST Circuits

Robert M. Bradley

CONTENTS

7.1 INTRODUCTION

In previous chapters, the current knowledge of the neurobiology of the mammalian brainstem gustatory relay nucleus has been detailed. Information presented shows that all chemosensory information derived from stimulating taste receptors, no matter where they are located, has to pass through the rostral nucleus of the solitary tract (rNST), and by the 1960s, the basic brainstem projection pattern of the afferent gustatory nerves had been established in outline form (summarized in Chapter 1, Figure 1.1). Details of the development of the connections are only now being studied and found to consist of complex overlapping terminal fields that suggest highly convergent input to the second-order neurons (Figure 6.2). Further anatomical pathway tracing mainly in rodents has established the projection patterns from the rNST to both rostral brain areas and brainstem sites (Figure 7.1). As described in Chapter 4, the brainstem connections from rNST are the secretomotor output to the salivary glands and motor output to various muscles involved in oral reflexes and facial expression. The rostral projection divides at the parabrachial nucleus, with one pathway passing through the thalamus to the

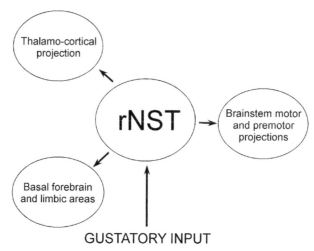

FIGURE 7.1 Diagram of the distribution pattern of gustatory afferent input after processing by the rNST.

cortex, whose function is believed to be involved in the sensory discriminative aspect of taste perception, whereas the other, limbic pathway is believed to be involved in the hedonic component of taste perception and its control of feeding. In addition, descending connections from forebrain areas to the rNST have also been described and to some extent investigated by examining the effects of electrical stimulation of these areas on the response characteristics of rNST neurons (Chapter 5).

Extracellular recordings were initially used to functionally establish that the rNST was in fact the brainstem taste relay, and then this technique was used by numerous later investigations to probe the role of the relay nucleus in taste processing. It is only in recent years that intracellular recordings have been used to characterize the biophysical and synaptic properties of rNST neurons.

Despite extensive investigations by many laboratories, the rNST essentially remains a "black box," and there is little information on what the rNST neurons do or how they interact as a neural circuit to process gustatory information. Nevertheless, investigators have made conclusions based on extracellular recordings from unidentified neurons in a limited part of the rNST on how the nucleus processes taste information. These experimental approaches have used similar techniques: Extracellular recordings are made from a sample of rNST neurons, and then a variety of ever more complex analyses applied to the data set. Often, only one concentration of a stimulus is used, and some feature of the resultant neural discharge is analyzed. It is apparent, therefore, though never stated, that several underlying assumptions are made in all these experiments that, if challenged, could undermine the conclusions that are drawn. It is my intent in this chapter to examine these assumptions more closely and suggest other ways to approach the study of the role of the rNST in processing taste stimulus-initiated neural activity.

7.2 KNOWLEDGE GAINED FROM EXTRACELLULAR RECORDINGS FROM RNST NEURONS

7.2.1 CENTRAL MAPPING OF ORAL RECEPTIVE FIELDS

Soon after the rNST was functionally defined by extracellular recordings of neuron responses to chemicals applied to the tongue,[1] the extent of the brainstem central gustatory projection was mapped.[2] This mapping study published in 1963 was remarkable for its time and consisted of sequential electrode penetrations made lateral to the midline at two planes, 2.9 and 1.5 mm rostral to the obex, and at different depths. Responses were recorded to taste, tactile, and thermal stimulations applied not only to intraoral receptive fields but also to the lateral face. The results were presented as a foldout map of the brainstem and provide the location in three coordinates of the chemosensory area of the rat brainstem. These pioneering findings were later expanded to determine if the gustatory relay had a chemotopic map in which the different taste qualities were localized in different dorso-ventral depths or areas of the rNST.[3,4] Although there were suggestions of variations in response magnitude with depth to different taste qualities, the existence of a chemotopic map similar to the tonotopic organization of the auditory brainstem nucleus was not found. However, later investigations described some level of a chemotopic organization.[5]

In investigations of the rostral-caudal response pattern to tongue stimulation, it is not surprising, given the anatomically defined projection pattern, to find that the electrophysiological responses to anterior tongue stimulation are maximum in the most rostral part of the rNST, whereas the posterior tongue chemosensitivity is recorded more caudally.[6–8] Caudal and medial to the rostral tongue area is an area that responds to both anterior and posterior tongue stimulation, suggesting interactions between anterior and posterior tongue receptive fields at the level of rNST.

Although early investigators restricted their studies to the projection pattern of the anterior and posterior tongue, the gustatory system in the rat is more widespread and, in addition to the anterior and posterior tongue, includes receptive fields on the anterior hard palate including the nasoincisor ducts, the soft palate (Geshmacksstreifen), buccal wall, sublingual organ, and epiglottis. In addition, the posterior tongue field consists of taste buds in both circumvallate and foliate papillae.[9,10] Taste buds in these areas are innervated by various branches of the VIIth, IXth, and Xth cranial nerves. Despite this widespread input to the rNST, most investigators have restricted analysis to the anterior tongue field principally because of expediency. Thus, much of the information on how the rNST processes taste information is based on recordings from only a small proportion of the peripheral gustatory receptors.

However, there are a few studies in which the response of rNST neurons to stimulation of the other gustatory receptive fields has been investigated.[11–14] It was discovered that rNST neurons have convergent input from the anterior tongue and nasoincisor duct taste buds, but the responses to taste stimuli applied to these two receptive fields differ. Apparently, the nasoincisor and soft palate taste buds are especially responsive to sucrose, whereas the anterior tongue taste buds are more responsive to NaCl. Some rNST neurons respond to whole-mouth stimulation but the majority respond to circumscribed stimulation of taste buds in one of the other receptive fields.[14] Furthermore, not all rNST neurons respond to gustatory stimulation; a significant proportion respond exclusively to tactile stimulation of various receptive fields in the oral cavity, suggesting that the designation of the rNST as a gustatory nucleus is somewhat of a misnomer. Even the gustatory responsive neurons are multimodal, responding as well to thermal and mechanical stimulation. In addition to differences in the response characteristics of the gustatory and mechanoresponsive neurons, they are located in different parts of the rNST: Gustatory responsive neurons are situated medial to the solitary tract, whereas the mechanoresponsive neurons are located lateral to the solitary tract. Thus, the rNST gustatory and mechanoreceptive neurons form parallel populations with nearly identical orotopic organizations, suggesting that in addition to being involved in sensory processes, these neurons provide information on location of ingested substances.

7.2.2 SOME BASIC ASSUMPTIONS DERIVED FROM THE EXTRACELLULAR RECORDINGS

Most investigators have used extracellular recordings of rNST neurons to investigate how taste information is encoded (Chapter 5). One of the underlying rationales for the early studies was to investigate if, at the level of the rNST, the relatively quality nonspecific afferent discharge patterns were somehow sorted out to become more specific. The hypothesis was that rNST neurons would be divided into sets of neurons more specifically responsive to the basic taste qualities. In one of the first investigations[15] of the transfer of information between afferent taste fibers and rNST neurons, the discharge characteristics to a single concentration of 14 taste stimuli recorded in chorda tympani fibers was compared to the response of rNST neurons to the same stimulus set. The results demonstrated that the characteristics of chorda tympani fibers were very similar to those of the rNST neurons with respect to response specificity. Changes did occur in response frequency, probably reflecting convergence between the afferent fibers and the second-order neurons.

Many later experiments used essentially the same animal preparation but employed different analyses (e.g., see References 16 and 17) to examine different characteristics of the neuronal discharge pattern. Usually, these investigators used a limited set of chemical stimuli, at single concentrations and at an undefined location of the rostral caudal extent of the rNST, although there are notable exceptions.[18] Mechanical and thermal sensitivity of the neurons was not usually determined. Also, because it is possible to record relatively few neurons per animal, in these experiments a large number of animals have to be used to obtain a suitable data set for analysis. Thus, the analysis relies on a limited sampling from each animal, but is analyzed as if the pooled set of neurons represents what would be recorded if all the neurons were sampled in a single animal. This becomes pertinent if the data are analyzed to look for cross-neuron discharge patterns as a way in which taste information is encoded.

At the time of the initial electrophysiological investigations of the rNST, little was known about the basic cell biology of the nucleus; electrodes were advanced into the rNST to isolate a neuron that responded to tongue stimulation with chemicals. Beginning in the mid-1980s, investigators began to describe the neuronal composition of the rNST.[19–21] As detailed in Chapter 2, neurons of the rNST have been divided into three morphological groups, one of which is comprised of interneurons. Beyond a general marker of the recording site (electrolytic lesion or dye mark), it is not possible using extracellular techniques to actually determine with any certainty the identity of the neuron that is being recorded from; thus, many of the electrophysiological recordings could have been derived from interneurons or other neurons with an undetermined projection pattern. However, the only definition of an rNST neuron is based on whether it responds to chemicals flowed over the tongue. These neurons are loosely called taste neurons, implying that they are involved in the perceptual properties of the sense of taste.

In a few later experiments, the projection pattern of the rNST neurons was determined by simultaneously electrically stimulating the pontine taste relay nucleus (PBN).[22–24] Thus, in a limited number of experiments, the recordings were obtained from neurons in the rNST that projected to the PBN, which presumably do participate in some aspect of the perceptual aspects of taste sensation, but does not rule out the possibility that these neurons also project or connect to neurons with other functions. In these studies, very few of the total population of rNST neurons were reported to project to the PBN (21–45%[22,23]). Anatomical studies, on the other hand, suggest a much higher number, although the counting was done at only two levels in the rNST and, therefore, did not represent a percentage of total rNST neurons.[25] In a more recent study of 101 rNST neurons that responded to tongue stimulation, 81 (82%) could be antidromically driven from the PBN.[24] However, the 101 neurons were obtained by recording from 73 animals, so that only a few rNST neurons were obtained per animal. The assumption is made that the total represents the results that would occur from recording from 101 neurons in a single animal. Differences in the results of these three studies relate to sampling problems inherent in the extracellular recordings. In the ideal experimental protocol, a large number of neurons would be sampled at both lateral-medial and rostral-caudal extents of the rNST, preferably in a single animal, to determine how the rNST processes information. The stimuli used should encompass several concentrations and include several members of the taste qualities as well as thermal and mechanical stimuli. It is also important to determine the identity and connections of the neurons being characterized. Current limitations of technology prevent such an experimental approach, although recent advances in the use of microwire and micromachined arrays of recording electrodes[26–28] may provide a promising new approach.

7.3 KNOWLEDGE GAINED FROM INTRACELLULAR RECORDINGS FROM rNST NEURONS

Intracellular recordings from neurons in the rNST would provide information on cellular morphology, projection pattern, synaptic characteristics, and biophysical properties that could not be achieved by extracellular recording. It is surprising, therefore, that of the hundreds of studies of the neurophysiology of the rNST, none were accomplished using intracellular recording. The only *in vivo* intracellular experiment used a technique in which extracellular recordings were made from rNST neurons to characterize chemical response characteristics, after which the cell was then penetrated with the same electrode and filled with a marker for later morphological analysis.[29] However, there is no direct confirmation that the extracellular recordings were obtained from the neuron that was subsequently filled, and no attempt was made after intracellular penetration to examine the biophysical properties of the neurons.

Intracellular recordings from rNST neurons were finally achieved using a brainstem slice preparation and provided data that could not be achieved by extracellular techniques.[30] The only drawback to this approach is that the brain slice is disconnected from the tongue, so that responses to taste stimuli cannot be recorded. However, much new information has resulted from these studies that contribute to a greater understanding of the neurobiology of the rNST.

7.3.1 *IN VITRO* RECORDINGS

Neurons of the rNST isolated from their input are not spontaneously active. Using a series of positive and negative current injections to determine basic membrane properties, rNST neurons were found to have a mean resting membrane potential of about –50 mV and an input resistance of 500 MΩ.[31] There is nothing unusual about these numbers, but it was possible to divide the rNST neurons into groups based on their repetitive discharge characteristics. Different neurons responded differently to a current injection protocol consisting of membrane hyperpolarization followed by a long depolarizing current injection. For some neurons, the initial hyperpolarization had no effect on the subsequent discharge pattern resulting from membrane depolarization. In contrast, other neurons responded to this current protocol by a delay in the initiation of action potentials (termed *delayed excitation*), and the length of the delay was related to either the length or magnitude of the hyperpolarizing prepulse (Figure 7.2).[32]

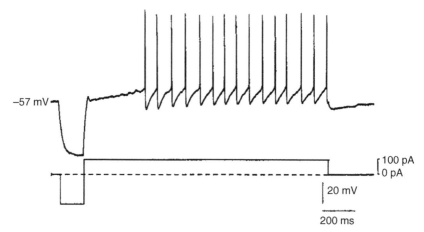

FIGURE 7.2 Intracellular recording from an rNST neuron. This neuron was briefly hyperpolarized and then depolarized by current injection (lower trace). The hyperpolarizing prepulse results in a long delay before action potentials are initiated by the depolarization.

The regular repetitive discharge pattern of a further group of neurons was changed to a highly irregular pattern by the hyperpolarizing prepulse, and the final group produced only a short burst of action potentials when depolarized. There was no correlation between the discharge pattern of the neurons and neuron morphology.[33] The ion channels underlying these different repetitive discharge patterns have also been investigated.[34] The variety of temporal firing patterns are important determinants of the way neurons transform synaptic input into spike output and, therefore, control information processing by the rNST. Importantly, afferent gustatory information has been shown to both excite and inhibit rNST neurons.[15,35] Excitatory input to rNST neurons would depolarize the membrane and initiate action potentials, whereas inhibitory input would result in membrane hyperpolarization. Thus, the neuron group unaffected by membrane hyperpolarization would pass on the afferent action potential train relatively unchanged. On the other hand, neurons with a delayed excitation would, depending on the combination of inputs, not pass on discharge pattern arriving from the primary afferent fiber input but change the pattern of neural discharges evoked in these neurons. These results demonstrated for the first time that rNST neuron groups could potentially process gustatory information differently.

Synaptic properties of rNST neurons have been detailed in Chapter 3. Besides identifying glutamate as the transmitter between the afferent input and all morphological types of rNST neurons, the role of inhibitory synaptic activity mediated by GABA is especially interesting. For example, the rNST neurons are chronically inhibited[36], repetitive stimulation results in long-term potentiation of inhibitory activity,[37,38] and all neuron morphological types respond to GABA.[39] The way that rNST neurons respond to trains of action potentials is a significant factor in how taste information is processed by the rNST and has been shown to influence the response properties of rNST neurons that respond to chemical stimulation of the tongue.[40] The role of the other neurotransmitters identified in the rNST (Chapter 3) in taste processing remains to be determined.

7.4 POSSIBLE CIRCUITS

7.4.1 rNST Circuits Based on Current Knowledge

In early experiments, investigators had a limited knowledge of the basic rNST circuit that processed chemosensory information. It consisted of an afferent input that synapsed with NST neurons that then project to more rostral brain areas.[1] Because response frequencies of rNST neurons are higher than afferent fibers in response to the same stimulus concentration applied to the tongue, it was assumed that convergence occurred at the first central synapse in the taste pathway (Figure 7.3). This conceptual model was the circuit underlying all the early experiments. Extracellular recordings were made from an rNST neuron in response to stimulation of taste buds with taste stimuli. With the discovery of projection patterns and different neuron morphologies, the basic circuit becomes more complex especially if input is derived from both the VIIth and IXth nerves (Figure 7.4). Two neuron types are illustrated in Figure 7.4. Projection neurons either connect to the PBN or synapse with other brainstem motor systems. The connections of the local circuit neurons have never

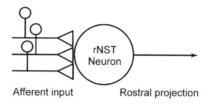

FIGURE 7.3 Basic circuit of the rNST that formed the basis of the early extracellular recordings. Convergent afferent input synapses on rNST neurons, which then project their output to rostral brain areas.

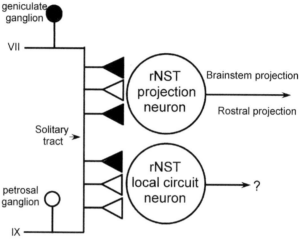

FIGURE 7.4 Circuit of the rNST that incorporates convergent input from both the VIIth and IXth nerves onto both projection and local circuit neurons. The projection neuron connects to either or both rostral brain areas as well as brainstem reflex connections. Filled black = projections from VIIth nerve; open = projection from IXth nerve.

been directly investigated, although anatomical and brain slice investigations have provided some details. They are known to be GABAergic[20,41] and are, therefore, presumed to synapse on the projection neurons to maintain tonic inhibitory activity.

Ultrastructural studies using GABA immunostaining and anterograde labeling of afferent terminals indicate that inhibitory synapses occur on the terminal endings, possibly derived from the inhibitory interneurons.[42,43] Stimulation of the solitary tract in brain slices initiates mixed postsynaptic potentials in rNST neurons. By blocking either the excitatory (glutamate) or inhibitory (GABA) component, it is possible to measure the latency between the stimulation pulse and the rise time of the synaptic potential. Excitatory postsynaptic potentials were found to have a mean latency of 4.8 ms and the inhibitory postsynaptic potential of 8.8 ms.[31] This difference in latency suggests that the excitatory postsynaptic potentials are mediated monosynaptically, whereas at least two synapses are involved in the inhibitory postsynaptic potential. Hence, the interneurons inhibit the afferent input by feedback inhibition (Figure 7.5).

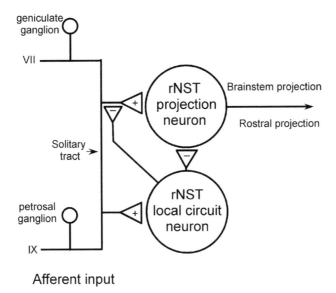

Afferent input

FIGURE 7.5 Circuit of the rNST that includes both excitatory and inhibitory synapses. Excitatory input to the local circuit neurons is diagrammed to inhibit the projection neuron, which is therefore chronically inhibited. As suggested by ultrastructural investigations, the local circuit neurons also reciprocally inhibit the excitatory afferent input.

7.4.2 MODEL rNST CIRCUITS

Neurons in the rNST receive input from afferent fibers innervating receptors in the oral cavity. The response of these rNST neurons is, therefore, determined by the response characteristics of the afferent fibers that synapse with the neuron. Despite recent advances that have suggested that taste quality may be coded using labeled lines,[44–46] most of the evidence based on recordings from single afferent gustatory fibers supports a population code.[47] Thus, peripheral gustatory fibers can be separated into groups according to greatest response to one of the taste qualities.[48] Recordings from rNST neurons reveal a similar grouping of responses (see Chapter 5).[16] During stimulation, not all the afferent fibers synapsing with a neuron will be activated equally because different fibers respond differently to particular taste qualities, so the response of an rNST neuron will reflect the overall level of activity in the afferent fibers.

At present, there is no information on the number of synapses between afferent chemosensory fibers and an rNST neuron. However, the receptive field size of rNST neurons is significantly larger than those of afferent taste fibers, demonstrating that afferent fibers innervating adjacent tongue-receptive fields converge on an rNST neuron.[49] Also, we do not know how many rNST neurons are synaptically connected to a single afferent fiber, although neural tracing studies show the terminal fields to be a complex meshwork of fibers. All of these factors will determine the response characteristics of an rNST neuron, as suggested in Figure 7.6. Moreover, little is known about the stability of these synaptic connections over time. The response characteristics of the afferent fibers may change as the taste receptor cells turn over.[50,51]

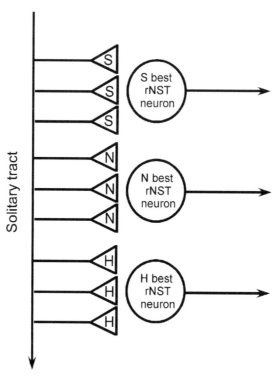

FIGURE 7.6 Circuit diagram that illustrates sorting of afferent fibers with similar response characteristics onto rNST neurons. rNST neurons will, therefore, have similar response characteristics to the afferent fibers. S = sucrose best fiber and rNST neuron, N = salt best fiber and rNST neuron, and H = acid best fiber and rNST neuron.

Because of this pattern of connections, the only way to alter the response profile of the rNST neurons is to change the afferent connections. This applies to all the neurons at each relay in the central taste pathway. It is not possible, for example, to modify a neuron at the PbN taste relay to change its response profile, because the connection pattern determines how the neuron responds to tongue stimulation.

Once the gustatory information has passed via the first central synapse to the second-order neuron in the taste pathway, it is uncertain what happens next, before the information is passed beyond the rNST. There is even the possibility that the rNST may be bypassed. For example, there is evidence that stimulation of the solitary tract can initiate postsynaptic potentials in some salivatory nucleus neurons with a latency of less than 4 ms, suggesting that this is a monosynaptic connection, indicating a direct reflex link between afferent input and the efferent output. The fact that the rNST contains a sizeable population of interneurons suggests processing, and as described above, inhibitory activity plays a significant role in rNST processing.

7.4.3 SEPARATION OF FUNCTION

Anatomical and electrophysiological studies reviewed above demonstrate that only a certain proportion of rNST neurons project rostrally to the PbN. This suggests that

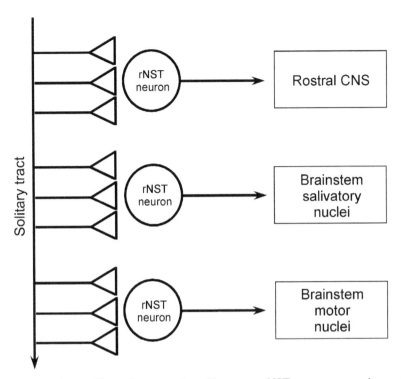

FIGURE 7.7 Diagram illustrating separation of input onto rNST neurons grouped according to projection pattern and function.

rNST neurons are separated into groups, some projecting rostrally, whereas others synapse with neurons involved in brainstem reflexes (Figure 7.7). Electrophysiological recordings from neurons that do not relay to the PbN revealed that they respond similarly to the relay neurons, albeit with some subtle differences.[23,24] Notably, rNST neurons responding best to sucrose preferentially project to the PbN.

Figure 7.7 is perhaps what many investigators assume happens to information arriving at the rNST. Converging afferent information arrives at an rNST neuron that is then distributed to rostral and brainstem sites. Conceptually, this circuit suggests that the rNST plays little or no role in sensory processing. Most of the processing takes place at the first synapse, where the rNST neuron averages all the sensory input and then sends it to various locations. Perhaps that is all that is necessary at this relay in the central taste pathway. However, because different taste qualities can result in different types and volumes of saliva, and different facial expressions, it would seem that some organization of the sensory input takes place in the rNST.

7.5 OUTLOOK FOR FUTURE INVESTIGATIONS

Based on the large body of literature reviewed in this book, there is a wealth of knowledge on the basic neurobiology of the rNST. The underlying rationale for all these experiments has been to understand how the rNST processes information derived from stimulating taste buds before it is distributed to other brain areas. Despite large gaps in knowledge on the basic connections of rNST neurons, investigators have not been deterred from drawing conclusions on rNST coding mechanisms. For example, much of the early analysis did not take into account inhibitory mechanisms or the influence of descending control of the rNST. Furthermore, sampling problems already alluded to contributed to other problems in interpretation.

Future investigations of the rNST require different approaches. It is important to be able to make ensemble recordings from the rNST. This can be accomplished with specially designed microelectrode arrays and will permit simultaneous recording experiments to determine how populations of neurons interact with chemical stimulation of the tongue. Use of imaging techniques on brain slices will also be useful to examine ensemble action of rNST neurons.[52,53] The use of differential interference contrast, infrared microscopy to visualize neurons identified with fluorescence, has also proved useful to record from known elements of rNST neural circuits.[54]

In future investigations, there is a need to define the circuits involved in processing by the rNST. Because a growing body of information suggests that different populations of rNST neurons exist that subserve different functional roles, it will be advantageous to identify and characterize these different populations. In my laboratory, we have already begun work on a relatively simple reflex circuit that connects the afferent sensory input to the parasympathetic secretomotor output.[55] Future investigations will use similar techniques to identify and characterize neurons with known projection patterns. Once identified, their synaptic characteristics and morphology will provide information on similarities and differences between neurons sending information rostrally and connecting locally in the brainstem. Further investigation of the rNST interneurons will also be of importance. Finally, although some ultrastructural studies have provided information on synaptic interactions of identified primary afferent taste fibers, more work needs to be accomplished on characterizing the degree of convergence, the number of rNST neurons contacted by a single afferent taste fiber, and differences in connections between primary afferent taste fibers and presumptive different functional rNST neuron groups.

It can be argued that until we have details of how rNST neurons are connected, it will be impossible to understand how the nucleus processes information. Extracellular recording and neuroanatomical studies have been enormously valuable for providing much basic information, but further progress needs different approaches. I have already made suggestions but other approaches are possible, such as using laser scanning photostimulation[56] and the endless possibilities available by using molecular genetic approaches. In appearance, the rNST seems to be a relatively simple nucleus and seems an ideal model system to study the connectivity of brain neurons. However, how rNST neurons connect and interact is the basis of all taste processing and holds secrets of central taste neurobiology.

ACKNOWLEDGMENT

The preparation of this chapter was supported in part by NIH grant DC 000288 from the National Institute on Deafness and Other Communication Disorders to R. M. Bradley.

REFERENCES

1. Pfaffmann, C., Erickson, R. P., Frommer, G. P. and Halpern, B. P. Gustatory discharges in the rat medulla and thalamus, in *Sensory Communication,* Rosenblith, W. A. (Ed.), MIT Press, Cambridge, MA, 1961.

2. Makous, W., Nord, S., Oakley, B. and Pfaffmann, C. The gustatory relay in the medulla, in *Olfaction and Taste,* Zotterman, Y. (Ed.), Pergamon Press, Oxford, U.K., 1963.

3. Halpern, B. P. Chemotopic organization in the bulbar gustatory relay of the rat, *Nature,* 208, 393, 1965.

4. Halpern, B. P. Chemotopic coding for sucrose and quinine hydrochloride in the nucleus of the fasciculus solitarius, in *Olfaction and Taste II,* Hayashi, T. (Ed.), Pergamon Press, New York, 1967.

5. McPheeters, M., Hettinger, T. P., Nuding, S. C., Savoy, L. D., Whitehead, M. C. and Frank, M. E. Taste-responsive neurons and their locations in the solitary nucleus of the hamster, *Neuroscience,* 34, 745, 1990.

6. Halpern, B. P. and Nelson, L. M. Bulbar gustatory responses to anterior and to posterior tongue stimulation in the rat, *Am. J. Physiol.,* 209, 105, 1965.

7. Ogawa, H. and Hayama, T. Receptive fields of solitario-parabrachial relay neurons responsive to natural stimulation of the oral cavity in rats, *Exp. Brain Res.,* 54, 359, 1984.

8. Sweazey, R. D. and Smith, D. V. Convergence onto hamster medullary taste neurons, *Brain Res.,* 408, 173, 1987.

9. Miller, I. J., Jr. Gustatory receptors of the palate, in *Food Intake and the Chemical Senses,* Katsuki, Y. et al. (Eds.), University of Tokyo Press, 1977.

10. Norgren, R., Nishijo, H. and Travers, S. P. Taste responses from the entire gustatory apparatus, *Ann. NY Acad. Sci.,* 575, 246, 1989.

11. Travers, S. P., Pfaffmann, C. and Norgren, R. Convergence of lingual and palatal gustatory neural activity in the nucleus of the solitary tract, *Brain Res.,* 365, 305, 1986.

12. Travers, S. P. and Norgren, R. Coding the sweet taste in the nucleus of the solitary tract: differential roles for anterior tongue and nasoincisor duct gustatory receptors in the rat, *J. Neurophysiol.,* 65, 1372, 1991.

13. Halsell, C. B., Travers, J. B. and Travers, S. P. Gustatory and tactile stimulation of the posterior tongue activate overlapping but distinctive regions within the nucleus of the solitary tract, *Brain Res.,* 632, 161, 1993.

14. Travers, S. P. and Norgren, R. Organization of orosensory responses in the nucleus of the solitary tract of the rat, *J. Neurophysiol.,* 73, 2144, 1995.

15. Doetsch, G. S. and Erickson, R. P. Synaptic processing of taste-quality information in the nucleus tractus solitarius of the rat, *J. Neurophysiol.,* 33, 490, 1970.

16. Smith, D. V., Van Buskirk, R. L., Travers, J. B. and Bieber, S. L. Gustatory neuron types in hamster brain stem, *J. Neurophysiol.,* 50, 522, 1983.

17. Di Lorenzo, P. M. and Victor, J. D. Taste response variability and temporal coding in the nucleus of the solitary tract of the rat, *J. Neurophysiol.,* 90, 1418, 2003.

18. Ganchrow, J. R. and Erickson, R. P. Neural correlates of gustatory intensity and quality, *J. Neurophysiol.,* 33, 768, 1970.

19. Davis, B. J. and Jang, T. The gustatory zone of the nucleus of the solitary tract in hamster: light microscopic morphometric studies, *Chem. Senses,* 11, 213, 1986.

20. Lasiter, P. S. and Kachele, D. L. Organization of GABA and GABA-transaminase containing neurons in the gustatory zone of the nucleus of the solitary tract, *Brain Res. Bull.,* 21, 623, 1988.

21. Whitehead, M. C. Neuronal architecture of the nucleus of the solitary tract in the hamster, *J. Comp. Neurol.,* 276, 547, 1988.

22. Ogawa, H. and Kaisaku, J. Physiological characteristics of the solitario-parabrachial relay neurons with tongue afferent inputs in rats, *Exp. Brain Res.*, 48, 362, 1982.

23. Monroe, S. and Di Lorenzo, P. M. Taste responses in neurons in the nucleus of the solitary tract that do and do not project to the parabrachial pons, *J. Neurophysiol.*, 74, 249, 1995.

24. Cho, Y. K., Li, C.-S. and Smith, D. V. Gustatory projections from the nucleus of the solitary tract to the parabrachial nuclei in the hamster, *Chem. Senses*, 27, 81, 2002.

25. Halsell, C. B., Travers, S. P. and Travers, J. B. Ascending and descending projections from the rostral nucleus of the solitary tract originate from separate neuronal populations, *Neuroscience*, 72, 185, 1996.

26. Katz, D. B., Simon, S. A. and Nicolelis, M. A. Taste-specific neuronal ensembles in the gustatory cortex of awake rats, *J. Neurosci.*, 22, 1850, 2002.

27. Nordhausen, C. T., Maynard, E. M. and Normann, R. A. Single unit recording capabilities of a 100 microelectrode array, *Brain Res.*, 726, 129, 1996.

28. Lehmkuhle, M. J., Normann, R. A. and Maynard, E. M. High-resolution analysis of the spatio-temporal activity patterns in rat olfactory bulb evoked by enantiomer odors, *Chem. Senses*, 28, 499, 2003.

29. Renehan, W. E., Jin, Z., Zhang, X. and Schweitzer, L. Structure and function of gustatory neurons in the nucleus of the solitary tract: II. Relationships between neuronal morphology and physiology, *J. Comp. Neurol.*, 367, 205, 1996.

30. Bradley, R. M. and Sweazey, R. D. Intrinsic characteristics of gustatory neurons in rat solitary nucleus. *Soc. Neurosci. Abstracts,* 15, 930, 1989.

31. Grabauskas, G. and Bradley, R. M. Synaptic interactions due to convergent input from gustatory afferent fibers in the rostral nucleus of the solitary tract, *J. Neurophysiol.*, 76, 2919, 1996.

32. Bradley, R. M. and Sweazey, R. D. Separation of neuron types in the gustatory zone of the nucleus tractus solitarii based on intrinsic firing properties, *J. Neurophysiol.*, 67, 1659, 1992.

33. King, M. S. and Bradley, R. M. Relationship between structure and function of neurons in the rat rostral nucleus tractus solitarii, *J. Comp. Neurol.*, 344, 50, 1994.

34. Tell, F. and Bradley, R. M. Whole-cell analysis of ionic currents underlying the firing pattern of neurons in the gustatory zone of the nucleus tractus solitarii, *J. Neurophysiol.*, 71, 479, 1994.

35. Travers, J. B. and Smith, D. V. Gustatory sensitivities in neurons of the hamster nucleus tractus solitarius, *Sens. Process.*, 3, 1, 1979.

36. Smith, D. V. and Li, C.-S. Tonic GABAergic inhibition of taste-responsive neurons in the nucleus of the solitary tract, *Chem. Senses*, 23, 159, 1998.

37. Grabauskas, G. and Bradley, R. M. Tetanic stimulation induces short-term potentiation of inhibitory synaptic activity in the rostral nucleus of the solitary tract, *J. Neurophysiol.*, 79, 595, 1998.

38. Grabauskas, G. and Bradley, R. M. Potentiation of GABAergic synaptic transmission in the rostral nucleus of the solitary tract, *Neuroscience*, 94, 1173, 1999.

39. Wang, L. and Bradley, R. M. Influence of GABA on neurons of the gustatory zone of the rat nucleus of the solitary tract, *Brain Res.*, 616, 144, 1993.

40. Lemon, C. H. and Di Lorenzo, P. M. Effects of electrical stimulation of the chorda tympani nerve on taste responses in the nucleus of the solitary tract, *J. Neurophysiol.*, 88, 2477, 2002.

41. Davis, B. J. GABA-like immunoreactivity in the gustatory zone of the nucleus of the solitary tract in the hamster: light- and electron-microscopic studies, *Brain Res. Bull.*, 30, 69, 1993.

42. Whitehead, M. C. Anatomy of the gustatory system in the hamster: synaptology of facial afferent terminals in the solitary nucleus, *J. Comp. Neurol.*, 244, 72, 1986.

43. Leonard, N. L., Renahan, W. E. and Schweitzer, L. Structure and function of gustatory neurons in the nucleus of the solitary tract. IV. The morphology and synaptology of GABA-immunoreactive terminals, *Neuroscience*, 92, 151, 1999.

44. Zhang, Y., Hoon, M. A., Chandrashekar, J., Mueller, K. L., Cook, B., Wu, D., Zuker, C. S. and Ryba, N. J. P. Coding of sweet, bitter, and umami tastes: different receptor cells sharing the same signaling pathways, *Cell*, 112, 293, 2003.

45. Scott, K. The sweet and bitter of mammalian taste, *Curr. Opin. Neurobiol.*, 14, 423, 2004.

46. Mueller, K. L., Hoon, M. A., Erlenbach, I., Chandrashekar, J., Zuker, C. S. and Ryba, N. J. P. The receptors and coding logic for bitter taste, *Nature*, 434, 225, 2005.

47. Smith, D. V. and Scott, T. R. Gustatory neural coding, in *Handbook of Olfaction and Gustation,* Doty, R. L. (Ed.), Marcel Dekker, New York, 2003.

48. Frank, M. An analysis of hamster afferent taste nerve response functions, *J. Gen. Physiol.*, 61, 588, 1973.

49. Vogt, M. B. and Mistretta, C. M. Convergence in mammalian nucleus of solitary tract during development and functional differentiation of salt taste circuits, *J. Neurosci.*, 10, 3148, 1990.

50. Beidler, L. M. and Smallman, R. L. Renewal of cells within taste buds, *J. Cell Biol.*, 27, 263, 1965.

51. Shimatani, Y., Nikles, S. A., Najafi, K. and Bradley, R. M. Long-term recording from afferent taste fibers, *Physiol. Behav.*, 80, 309, 2003.

52. Sato, K. and Momose-Sato, Y. Optical mapping reveals developmental dynamics of Mg^{2+}-/APV-sensitive components of glossopharyngeal glutamatergic EPSPs in the embryonic chick NTS, *J. Neurophysiol.*, 92, 2538, 2004.

53. Baker, B. J., Kosmidis, E. K., Vucinic, D., Faulk, C. X., Cohen, L. B., Djurisic, M. and Zecevic, D. Imaging brain activity with voltage- and calcium-sensitive dyes, *Cell Mol. Neurobiol.*, 25, 245, 2005.

54. Fukami, H. and Bradley, R. M. Biophysical and morphological properties of parasympathetic neurons controlling the parotid and von Ebner salivary glands in rats, *J. Neurophysiol.*, 93, 678, 2005.

55. Bradley, R. M., Fukami, H. and Suwabe, T. Neurobiology of the gustatory-salivary reflex, *Chem. Senses*, 30, i70, 2005.

56. Sternson, S. M., Shepherd, G. M. G. and Friedman, J. M. Topographic mapping of VMH → arcuate nucleus microcircuits and their reorganization by fasting, *Nature Neurosci.*, 8, 1356, 2005.

Index